Cosmological Enigmas

Cosmological Enigmas

Pulsars, Quasars &
Other Deep-Space Questions

MARK KIDGER

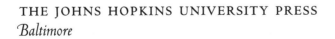

THE JOHNS HOPKINS UNIVERSITY PRESS
Baltimore

The Johns Hopkins University Press
2715 North Charles Street
Baltimore, Maryland 21218-4363
www.press.jhu.edu

Library of Congress Cataloging-in-Publication Data

Kidger, Mark R. (Mark Richard), 1960–
    Cosmological enigmas : pulsars, quasars, and other deep-space questions /
Mark Kidger.
        p.   cm.
    Includes bibliographical references and index.
    ISBN-13: 978-0-8018-8460-3 (hardcover : acid-free paper)
    ISBN-10: 0-8018-8460-8 (hardcover : acid-free paper)
    1. Cosmology—Popular works.   I. Title.
    QB982.K53 2007
    523.1—dc22   2007014811

A catalog record for this book is available from the British Library.

Page 225 constitutes an extension of this copyright page.

Special discounts are available for bulk purchases of this book. For more information, please contact Special Sales at 410-516-6936 or specialsales@press.jhu.edu.

To Sir Patrick Moore

For opening the author's eyes to the wonders of the Universe

as a young child with his BBC program *The Sky at Night*

Contents

Acknowledgments   *ix*

Introduction   1

CHAPTER 1
How Are Stars Born and How Do They Die?   5

CHAPTER 2
How Do We Know That Black Holes Exist?   24

CHAPTER 3
Who Is the Strangest in the Cosmic Zoo?   39

CHAPTER 4
How Far Is It to the Stars and Will We Ever Be Able
to Travel to Them?   61

CHAPTER 5
How Old Is the Universe?   80

CHAPTER 6
Is Anybody There? 96

CHAPTER 7
How Will the Universe End? 117

CHAPTER 8
Why Is the Sky Dark at Night? 136

CHAPTER 9
How Do We Know There Was a Big Bang? 154

CHAPTER 10
What Is There Outside the Universe? 174

Notes 193
Index 217

Color galleries follow pages 84 and 116.

Acknowledgments

Many people have contributed directly or indirectly to this book and its sister volume. In most cases, the help has been anonymous or "E Pluribus"; in other cases, it has been so direct and important that it would be churlish not to acknowledge it. The team at Johns Hopkins University Press has been extremely supportive and professional, starting with my editor, Trevor Lipscombe. Most people do not realize how important an editor's role is in a successful book project: ideas, encouragement, suggestions, corrections, and modifications all have come from Trevor and his encyclopedic knowledge. Other staff at JHUP, in particular Juliana McCarthy and Bronwyn Madeo, have also gone well above and beyond the call of duty. I have also been privileged to work with two eagle-eyed copy editors who have done far more than just correct spelling mistakes and bad grammar. Finally, a vote of thanks to supportive colleagues, both former ones in Tenerife and the amazing Herschel team in Madrid—the best team in the world.

Cosmological Enigmas

Introduction

Why study the Universe? A few years ago the president of the British Astronomical Association entitled his presidential lecture, "Let's Be Useless," a whimsical nod to the many people who think that astronomy is a harmless, entertaining, but totally useless pursuit. What is the point in studying the Universe when millions in the world are starving? Should we be spending huge sums on new telescopes to study the distant confines of the Universe, or could that money be better used close to home? This book does not try to answer directly why we study the Universe. Instead of justifying *why* we study, its purpose is to explain *how* we find out the things we think we know. With such information, the reader may (or may not!) be persuaded that my fellow astronomers and myself, far from being a useless luxury, actually do serve some purpose.

Some time after this volume is published, the Spanish Gran Telescopio CANARIAS (Canaries Large Telescope, usually known as the GTC) will enter service. It will be the second-largest single-mirror telescope in the world with a largest diameter of 11.4 meters,[1] and a surface area equivalent to a full 10-meter diameter, surpassing the two Keck telescopes atop Mauna Kea in Hawai'i. This monster is probably the largest scientific project that Spain has ever undertaken and comes with an impressive price tag that, when the project was approved, was estimated at 13 billion pesetas (about $90 million). This may sound like a lot of money, but it is less than a third of the cost of a new A380 airliner, and about the same as a new Typhoon Eurofighter, or two (now obsolete) F-14 Tomcats. The GTC has cost the average Spaniard about 30 cents a year over the course of the project, less than a quarter of what the average Spaniard spends on midmorning coffee. The cost of astronomy, even the large, ambitious projects that are sometimes called "big astronomy," is a lot less than most people think.

What do you get for your dollars? Part of the answer is technology: solving scientific and engineering problems almost always turns out to be profitable in the end because it challenges industry to test itself and find its limits. This new technology may later find hundreds of unexpected uses (telescope construction, for example, poses big challenges in precision engineering, electronics, and optics, to say nothing of computing and heavy engineering). More than anything, though, what we are buying is knowledge, and that is something that will never go away. A navigator called Christopher Columbus had this crazy idea that you can cut the time taken to travel to the Indies by going west instead of struggling around Africa to get there. He hawked his idea around the royal courts of Italy, Spain, and Portugal, where most people just laughed at the idea because they *knew* that the Earth was flat and that, by sailing west, Columbus would fall off the edge of the world. Should he have given up because his voyage seemed silly and irrelevant? Just imagine what would have happened had Columbus lacked the temerity to persist with his "ridiculous" question about what would happen if he sailed west and not east.

We learn by asking ridiculous questions and challenging existing knowledge and facts. For 1,000 years after the fall of Rome, the human race stopped asking questions, seeking answers, and pushing itself to new heights; the result of this period of stagnation was the Dark Ages and later the Middle Ages, when the quality of human life dropped abysmally. Sanitation, medicine, communications, and culture in general dropped to a frightening degree, and most people lived in misery—even the richest and best-off had a poor diet and poor sanitation and, as a result, were prone to diseases. The Black Death, for example, eradicated perhaps one-third of the entire population of Europe (in Venice 60 percent of the popu-

lation died in just 18 months). I am not saying that the Black Death would have been avoided if our ancestors had built a few 10-meter telescopes, but had they retained even a fraction of the accumulated knowledge of the Greeks and Romans, an awful lot of misery could have been avoided.

Scientific knowledge, even abstract knowledge, is a part of human culture, and its accumulation over 4 million years of human evolution fuels our advance. Its beginnings have been immortalized in the scene in *2001: A Space Odyssey* in which the man-ape Moon-watcher discovered that using a bone as a weapon lengthened his reach, strengthened his blow, allowed him to feed his family better, and thus avoid the extinction of the line that would lead to *Homo sapiens*. When Moon-watcher throws the bone into the air in triumph it changes into a spacecraft; Stanley Kubrick shows, brilliantly in this flash forward, how a tiny discovery for a man-ape would eventually lead to modern man and spaceflight.

With each grain of knowledge added to our store, we have advanced a little further along the evolutionary course. What appeared to be abstract and useless research on "animal electricity" by the English bookbinder Michael Faraday in the early nineteenth century led to the development of the dynamo. When combined with the independent discovery that crude oil, when refined, gives highly combustible products, dynamos could power motors, which in turn would bring cheap light and power to the entire world. Scientific research has many blind alleys and dead ends, but even the most abstract research can pay off in a way that could never be imagined. The peaceful uses of radioactivity in medicine and engineering, for example, would have struck its nineteenth-century discoverers almost as witchcraft. Research to counteract a possible Nazi death ray led, directly, to radar (making air traffic control and thus mass transport by air possible) and also to high-capacity long-distance communications. It even made microwave ovens possible.

While attempting to explain electricity to the then prime minister Sir Robert Peel, Faraday was asked, "What good is it?" Faraday replied with feeling, "What good is a new-born baby?" According to the legend, Faraday added, "Rest assured, one day you will tax it." Numerous attempts have been made by different governments to fund only "useful" practical research; these have been doomed to failure and have rarely produced great results, while cutting off other lines of progress. One classic case was the Nixon administration's aim to find a cure for cancer by transferring an important part of the funding that NASA had received for the Moon landings to this effort; no great breakthrough was found. Although improved funding and a big research effort will often help solve a problem, large sums of money and practical research do not guarantee results—in fact, the kind of lateral thinking that apparently abstract research implies is, surprisingly often, more cost effective in the long run than applied research.

The greatest reason for doing abstract research is really totally different, however. When we study distant quasars and galaxies, exotic black holes and strange stars, we satisfy a different hunger and thirst: a thirst for knowledge and a hunger to know the answers. We are a restless race, and if we do not expand our mental horizons, we stagnate and wither. One hundred years ago many inhabitants of the Canary Islands dreamed of a distant land and of one day escaping to it; some made it to South America (and particularly Venezuela), although many died in the attempt (the police had orders to shoot to kill people trying to reach ships and emigrate illegally). Today our horizons are as likely to be distant planets as distant countries—crime rates round the world dropped dramatically the night of the first Moon landing, showing just how much our mental horizons have expanded. Our news bulletins alternate images of wars with those of distant galaxies photographed by the Hubble Space Telescope, and scenes of politicians are followed by panoramas of Mars.

Traveling around the world talking to astronomers and members of the public alike, I find the degree of fascination that people have with black holes, quasars, and life on other planets is unlimited. Occasionally, I take part in a BBC World Service phone-in program for South America where members of the public put their questions to a scientist; the range and level of questions never ceases to stagger me, and, on occasion, it is hard to give simple answers that are suitable for broadcast. People want to learn more about news items. They seek to understand complex questions of astrophysics because the Universe around us has—dare I say it—a universal appeal and fascination. For the past 50 years no corner of the Earth has remained unexplored; the great adventurers of the twenty-first century are those select few who probe the distant confines of the Universe, finding new marvels that help us to understand how wonderful it is. With each new discovery we also discover more about ourselves, our place in the Universe, the incredible beauty of the cosmos, and how small and delicate our Earth is. Our telescopes reveal ceaseless wonders around us, wonders that we are now able to understand to a greater or lesser degree. This knowledge both fascinates us and enriches us, and our desire to know and understand these marvels makes us truly *Homo sapiens*.

CHAPTER I

How Are Stars Born and How Do They Die?

Twinkle twinkle little star
Now we know just what you are
Nuclear furnace in the sky
You'll burn to ashes by and by

When we see stars twinkling up in the sky, little do we think or even imagine that each one is a huge sun. In fact, almost all of the 2,500 to 3,000 stars that we can see on any clear night with the naked eye from a dark location are bigger and more luminous than our Sun. It is a mind-boggling demonstration of our Sun's lack of importance.

The idea that the stars were just distant suns is an ancient one. It seems that the idea was suggested in ancient Greece; certainly the ideas of the philosopher Xenophanes of the sixth century B.C. suggest that he believed that there were many suns (although in other senses his ideas were much less advanced than those of his contemporaries). Greek philosophers fell broadly into two schools: one believed that the Earth was flat and that the heavens rotated around it (Heraclitus of Ephesus suggested in the sixth century B.C. that the diameter of the Sun was about 30 centimeters); the other believed the Earth was spherical and rotating and that it moved around the Sun. Amazingly, as far back as the fifth century B.C. Anaxagoras of Clazomenae suggested that the Sun was a glowing orb. He suggested that it was a giant red-hot stone and that the Moon shone by reflected sunlight. This seems to have been the first serious attempt to explain the nature of the Sun. By the fourth century B.C. Heraclides of Pontus had suggested that the Earth was rotating—thus explaining the rotation of the heavens—and that Mercury and Venus revolve around the Sun and not the Earth. In the third century B.C. Aristarchus of Samos suggested that the Earth moves around the Sun and even estimated the relative distances of the Sun and the Earth. Greek thought came to its apogee with Eratosthenes of Cyrene, the librarian at Alexandria in the third century B.C., who made an accurate measurement of the size of the Earth.

From the third century B.C. Greek thought went backward. The advanced ideas of the philosophers of the fifth to third century B.C. were forgotten and the idea of a flat Earth in the center of a crystal sphere on which the planets and the stars were supported held sway. It would take astronomers 1,800 years to start to get back to where they had been in the third century B.C. and another century to throw out completely the idea that the Earth was the center of the Universe with everything revolving around it.

Strangely, though, few philosophers apart from Anaxagoras wanted to speculate on the nature of the Sun and the stars. For most astronomers, they were just sources of light in the heavens whose nature was a mystery. Not until the nineteenth century would astronomers start to think seriously about the nature of stars and not until well into the twentieth century would they start to understand how stars are born and how they die.

Even now there are many things we do not understand well, particularly about stellar deaths. We are beginning, though, to get a good idea of how the stellar family came to be and how the stars live and die. The story of how astronomers have tried to explain the stars and the reasoning used is a fascinating tale of logic and missed opportunities.

Historical Errors and Successes

In 1838 the German astronomer Friedrich Bessell became the first person to measure the distance to a star. He showed that the fifth magnitude star 61 Cygni was slightly more than 11 light years or 95 million million kilometers away.[1] At the same time, Friedrich George Struve, a German astronomer working in St. Petersburg, and the Scottish astronomer Thomas Henderson,[2] working at the Cape Observatory in South Africa, measured the distance to the brilliant stars Vega and Alpha Centauri, respectively. The results were captivating. Bessell had elected 61 Cygni because it was a double star. Its two components were widely separated, suggesting that they must be relatively close to Earth. Henderson picked the third brightest star in the sky—also a widely separated double. Struve picked Vega because it was also extremely bright and passed almost overhead from St. Petersburg, making it easy to observe. It turned out that, although Alpha Centauri and Vega were similar in apparent brightness, their distances were not. Alpha Centauri was only 4.3 light years away, whereas Vega was 26 light years from Earth.

These results made it clear that, although apparently the same brightness, the stars are really at quite different distances and so must be of very different luminosities. In addition, the stars were of different colors and thus had different temperatures. It was well known that a bar of metal, when heated, would first start to glow dull red and then, as the temperature increased, would pass through orange, yellow, and white, going through the colors of the rainbow, before finally melting. Thus Vega, being white, was obviously a hotter star than 61 Cygni, which was orange.

By their choice of stars of different colors, Henderson, Bessell, and Struve, without knowing it, had laid down the bases of one of the most important clues to the nature of the stars themselves. You can see it for yourself in table 1.1. The cooler the star, the less luminous it is. Vega, blazing white hot, is 50 times more luminous than the Sun, but the Sun is itself 15 times more luminous than the orange star 61 Cygni. Without even knowing what powered the stars, astronomers could have known that the hotter a star was, the more luminous it was. They were also able to work out another factor: the difference in luminosity between Sirius and 61 Cygni was far too great to be explained simply by the lower temperature of 61 Cygni. An orange star emits less light than a white one because it is cooler, but not

Table 1.1   Color and Luminosity of the Stars
Measured by Struve, Henderson, and Bessell

Star	Color	Luminosity (Sun = 1)
Vega	White	50
Alpha Centauri	Yellow	1.4
61 Cygni	Orange	0.07

so much less as to explain the difference in luminosity. The answer was simple: the cooler a star, the smaller it is. Not only does the surface of a star emit less light if it is cool, but there is a smaller surface area of the star to emit light. So, as stars get cooler, their luminosity plummets. In contrast, hotter stars are far more luminous. By 1840 astronomers had enough information to uncover one of the fundamental laws that govern stellar physics; however, another 70 years would pass before someone looked at the data in the right way and asked the right question.

In the meantime, astronomers started to use a powerful new tool to study the heavens. Back in 1665 Isaac Newton carried out a series of experiments splitting up sunlight using a prism.[3] He established that light could be broken up into the colors of the rainbow to form a spectrum. At the time this seemed little more than a curiosity. Two discoveries based on Newton's experiments but made more than a century later were to have profound implications for astronomy in the twentieth century. First, in 1800 William Herschel discovered that a thermometer warmed up when placed at different points inside the solar spectrum. When he placed the thermometer beyond the end of the visible spectrum, he discovered that the thermometer continued to warm up. In this way he demonstrated the existence of infrared light—light beyond the red—which is now such an important tool of modern astronomy.[4] Herschel was also the first person to examine the distribution of the light in the spectra of stars of different colors and note the profound differences that would later be the basis of spectral classification, one of the most powerful tools of the modern astronomer. Similarly, in 1802 the English astronomer William Woolaston noticed the presence of dark lines crossing the Sun's spectrum. Initially these were not regarded as being particularly significant; a popular theory was that they were no more than the boundaries between different colors. By

1815, though, the German astronomer Joseph von Fraunhofer had produced a detailed atlas of the solar spectrum showing no fewer than 324 dark lines crossing it. These lines came to be known as Fraunhofer lines.

What the Fraunhofer lines were remained a mystery until 1859. In that year, two German astronomers, Gustav Kirchoff and Robert Bunsen,[5] explained how these lines were formed. They showed that when different elements are heated in a flame and the light is examined with a spectroscope, each element produces a characteristic set of bright lines. The lines from each element are as individual as a fingerprint. When superimposed on a bright rainbow spectrum, however, the lines appear dark. In the Sun, the dark lines are created by slightly cooler gas that lies above the bright solar photosphere and absorbs its light. By looking at the dark lines in the solar spectrum and comparing them with the lines produced by different elements in the laboratory, we can examine the composition of the Sun. When astronomers did this, they became aware of an orange-yellow line that could not be explained by any known element. This element was christened *helium* (from the Greek word *helios*, the Sun), by the English astronomer Sir Norman Lockyer, who discovered it in 1869. Eventually helium was identified as the gas produced by certain radioactive elements on Earth. Helium is the only element discovered in an astronomical object before it was found on Earth. Its discovery demonstrated the power of spectroscopy to help astronomers to understand the stars.

By the 1860s astronomers were starting to examine the spectrum not just of the Sun but of many stars and even some of the brighter nebulae. They recognized that stars had very different types of spectra according to their color. In 1863 Angelo Secchi made the first systematic attempt to classify stellar spectra using more than 500 stars. He noted that the hottest stars, the blue and white ones like Sirius or Vega, had very strong broad, dark hydrogen lines but showed little evidence of metals. Yellow stars like the Sun still showed hydrogen lines, but they were much less prominent. In contrast, however, the metals were much stronger. Orange stars showed much more complicated spectra with many bands—broad, dark features rather than individual lines. And the coolest and reddest stars had many broad carbon lines.

Because Secchi's classification of stars was too simple, it soon became obvious that something better was needed. That something was provided by the director of Harvard Observatory, Edward Pickering. Pickering realized that there were more subtleties in stellar spectra than Secchi had seen with his primitive equipment. Pickering called the normal white stars type A. A class of hot, blue-white stars showed somewhat weaker hydrogen lines and stronger helium lines. The ones with slightly weaker lines were type B. The types then went on through C, D, E, F, through to type M, the coolest and reddest stars. This system of classifying stellar spectra, developed by two staff members at Harvard, Annie Jump Cannon and Wilhelmina Fleming, became known as the Harvard system.[6]

Cannon and Fleming found that the original A, B, C, . . . system had numerous errors. Type B stars were actually hotter than type A stars, so their order was inverted. Other types, such as C, D, and E were duplications. In the end, the sequence became B, A, F, G, K, M. Two very rare classes of extremely hot stars were found and termed W and O. At the red end, some extra classes representing rare, extremely cool stars were added: R, N, and S. Class P was added for gaseous nebulae and Q for novae, but these classes are now rarely if ever used. For the stars that remained, the simple alphabetical list became W, O, B, A, F, G, K, M, R, N, S. Astronomers remember the sequence using the mnemonic "Wow! Oh Be A Fine Girl Kiss Me [Right Now Smack]."

As has happened many times in the history of astronomy, a major advance that in retrospect was blindingly obvious was eventually made almost simultaneously and independently by two people. In Denmark, Ejnar Hertzsprung had the idea in 1911 of representing the absolute magnitude of stars—this is, the magnitude that the star would have if it were at a standard distance of exactly 32.6 light years[7]—against their spectral type using the new Harvard system. The American astronomer Henry Norris Russell had the same idea in 1913. The result was the Hertzsprung-Russell diagram, usually abbreviated to H-R diagram, one of the fundamental tools of modern astrophysics. One of the earliest examples is shown in figure 1.1. The same diagram could have been prepared 50 years earlier using the colors of stars—blue, white, yellow, orange, red, and their graduations—and would have given a similar result, as we can see from table 1.1.

The results were striking. Most stars lay in a broad diagonal band from the hottest stars in the top left of figure 1.1, to the coolest and least luminous in the bottom right. These dim red stars were evidently much smaller than the Sun and were christened *red dwarfs* by astronomers. They are the tiny glow worms of the celestial bestiary. In contrast, the W and O stars are not only hot but are also extremely luminous. The surface of a typical type O star may be 40,000°C, far higher than the 6,000°C of the Sun, and may have 50,000 times the luminosity. In contrast, a red dwarf's surface is at about 3,000°C and may have one-ten-thousandth (or less) of the Sun's luminosity. So, the difference in temperature between the hottest and the coolest stars is a factor of 13, but their luminosity is different by a factor of 500 million.

The band in the H-R diagram became known as the Main Sequence. Most stars were spread along it. A few stars, though, were orange or red and very luminous and formed a group, shown in the top right corner of figure 1.1. For a star to be red and cool, but tens of thousands of times more luminous than the Sun, it has to be enormous. The Sun is 1,392,000 kilometers in diameter, but some of these red stars had to measure hundreds of millions of kilometers in diameter for the numbers to fit. These stars became known as red giant and supergiant stars.

The H-R diagram also showed that there were stars connecting the region of the red giants and supergiants with the Main Sequence, which provided astronomers with a false clue about the formation of stars. It made them think that the band of stars that they were seeing was an evolutionary sequence (it was, but not in the way that they thought).

Glowing Coals and Shrinking Clouds

Once astronomers knew how distant and luminous stars were, they began to speculate more clearly on their nature. In his classic book *The Scenery of the Heavens,* published in 1890, the English astronomer John Ellard Gore summarized knowledge of the heavens in the late nineteenth century. He described recent advances in spectroscopy and in stellar studies but did not broach the subject of the nature of stars. The first plausible theory of how the Sun obtains its energy, though, had been proposed as early as 1853.

The German astronomer Robert Mayer demonstrated in 1849 that the Sun could not be an inert, glowing ball of gas because it would cool down in just 5,000 years. Nor could it burn in a conventional sense, for even if it were made of coal, it would burn out in 4,600 years. Both estimates were soon rejected, because by then geological evidence showed that the Earth was at least millions of years old. This demonstrated that the Sun had some form of continuous generation of energy far more efficient than burning. Mayer proposed that the energy was generated by the impact of meteorites falling onto the solar surface. This scenario was implausible; to work, it required the Sun's mass to double every 30 million years.

A far more plausible theory was proposed in 1853 by Hermann von Helmholtz. He calculated that if the Sun's diameter were shrinking progressively by just 60 meters per year, this gravitational contraction would produce all the energy that is observed, and this energy supply would last for 15 million years. However, Lord Kelvin, the great British physicist of the late nineteenth century, pointed out the discrepancy between the age of the Sun on this theory and the age of the Earth from geological evidence.

The H-R diagram, though, lent support to the gravitational contraction theory. It seemed reasonable that a star could start as a huge, inert cloud of gas that would contract under its own force of gravity. Henry Russell himself suggested that the H-R diagram represented the evolution of stars from such a cloud. As it contracted, it would start to glow, first a dull red, before getting smaller and hotter. A star would pass through red, orange, and yellow to white, each time getting hotter and shrinking further. When its luminosity peaked, it would then slowly start to cool, shrinking and sliding down the Main Sequence until it ended up as a dim red dwarf. In other words, a star would start as a red supergiant, pass through being a red giant, and then become an orange giant star until reaching the Main Sequence. By this theory, the Sun would be a rather old star approaching the end of its life.

It was not astronomers, though, but geologists who finally demonstrated that this idea was untenable, at least in its original form that speculated that all the energy came from the gravitational collapse itself and no other intrinsic source. (The idea that, as a star aged, it would slowly travel down the Main Sequence, fading as it did so until ending as a dim red dwarf, held until the time of the Second World War.)

At the end of the nineteenth century, an energetic debate ensued about the Earth's age. The same arguments were applied as had been previously applied to the Sun. Physicists demonstrated that the Earth could not be more than a few million years old because the interior would have cooled and left the planet inert, which volcanic activity clearly shows that it is not. Geologists, though, looked at how long it took layers of rock to form and concluded that the age of the planet had to be hundreds or thousands of millions of years. There ensued a heated and sometimes bitter debate, with no flaw found in the logic of either camp. While it could be argued that the Earth was only a few million years old, the gravitational-contraction theory for generating the Sun's energy seemed plausible because it would make the Sun's age agree with the Earth's.

The key to resolving the dispute was the discovery of radioactivity. Both sides had presented indisputable facts. But what neither side knew at the time was that the decay of radioactive elements inside the Earth produces large quantities of heat. This energy is certainly enough to keep the interior of the Earth hot for thousands of millions of years. In particular, the large amounts of uranium and radioactive potassium in the Earth's interior generate the heat that was needed to make the books balance and explain the difference between the calculations of the physicists and the geologists.

With the problem of the age of the Earth solved, it seemed certain that the Sun was at least as old—about 5,000 million years old was the age calculated from the decay of radioactive elements—and possibly much older. What source of energy could power the Sun for so long, given that Lord Kelvin had shown that gravitational collapse would certainly do so for no longer than 50 million years?

Atoms for Energy

Legend has it that the key to the energy that powers the Sun and the stars was worked out in a train carriage in 1938. In fact, the theory had been proposed some years earlier. The distinguished British astronomer Sir James Jeans commented in 1930 that the American astronomer Charles Perrine and the British astronomer Sir Arthur Eddington had proposed that nuclear fusion—the building up of more complex elements from hydrogen—might be the explanation.[8] In 1926 Eddington suggested that converting hydrogen into helium could produce the Sun's energy. At the time, though, the theory was not widely accepted for a curious reason. Jeans pointed out that even converting the whole mass of the Sun from hydrogen to helium would only maintain the Sun for 100,000 million years, whereas there was circumstantial evidence that the stars are a million million (a U.S. trillion) years old. So, nuclear fusion could not be the answer either. He favored a hypothesis in which the Sun converted matter directly into energy.

He was largely correct in his estimate of the time that hydrogen fusion would last, but the age of the stars was massively overestimated.[9] We now know that the Sun is about the same age as the Earth.

Finally, with the storm clouds of war gathering over Europe and the Pacific, an exiled German physicist, Hans Bethe, who had been appointed an assistant professor at Cornell University in 1935,[10] was returning to Ithaca from Washington by train. Bethe was an expert on nuclear physics, and to pass the time on the journey, he started to examine the energy source of stars. By the end of the journey he had the theory worked out.

Bethe supposed that protons—hydrogen nuclei—in the Sun would combine with carbon nuclei in a chain of reactions that would give rise to the formation of helium. This is now known as the CNO cycle (carbon-nitrogen-oxygen) for the intermediate elements involved. In this cycle a proton combines with a carbon nucleus to form nitrogen. By decay and addition of further protons, the nucleus builds up in mass through different varieties of carbon and nitrogen to form oxygen. The oxygen breaks into the original carbon nucleus plus helium. Because the carbon is not consumed in the reactions and at the end remains unchanged, scientists refer to it as a catalyst—it makes the reactions possible but is not used up in them.

Technically, the CNO cycle, sometimes called the Bethe-Weizsäcker cycle,[11] is:

$$^{12}C + p \rightarrow {}^{13}N + g$$
$$^{13}N \rightarrow {}^{13}C + e^+ + n$$
$$^{13}C + p \rightarrow {}^{14}N + g$$
$$^{14}N + p \rightarrow {}^{15}O$$
$$^{15}O \rightarrow {}^{15}N + e^+ + n$$
$$^{15}N + p \rightarrow {}^{12}C + {}^4He$$

In the end, four protons (p) get added to the carbon nucleus to form helium. On the way, two positrons or antielectrons (e^+) and two gamma ray photons (g) are released. The helium nucleus has less mass than the four protons used to produce it. So, in effect, part of the mass of the hydrogen gets converted into energy.

The CNO chain is not the only way to generate energy. Because a nitrogen atom has six protons and thus a positive charge of six, it is difficult to combine it with a proton because like charges repel each other and a star needs a high temperature and pressure inside to force the two together. This only happens efficiently in stars larger and more massive than the Sun. Bethe and his colleague Charles Critchfield found an alternative hydrogen reaction called the proton-proton chain, usually abbreviated to the p-p chain:

Cosmological Enigmas

$$p + p \rightarrow d + e^+ + n$$
$$d + p \rightarrow {}^3He + g$$
$$He + {}^3He \rightarrow {}^4He + p + p$$

In this reaction chain two protons (p) combine to form a heavy hydrogen or deuterium nucleus (d). A proton is then added to this nucleus to form helium-3 or light helium. Finally, two light helium nuclei combine to form a normal helium nucleus. The end result is the same as in the CNO chain: four protons combine to form a helium nucleus, plus two gamma ray photons, two positrons, and two neutrinos. Bethe and Critchfield had discovered the force that powers the Sun.

A Star Is Born

Once astronomers understood the nuclear reactions that power stars—and thus how they live—they could think seriously about how stars are born and evolve. All around the sky we see nebulae. Observers noted that some appeared white in a large telescope, and some of these could even be split into stars—these we now know as galaxies. Others had a greenish color[12] and could not be split into stars even with a large telescope under high magnification—these are what we would now call *nebulae* from the Latin word for clouds. In the late nineteenth century, spectroscopic studies, most notably by Sir William Huggins in England, confirmed what visual observers had suspected: that the green nebulae are glowing clouds of gas. In many cases these green nebulae seemed to be associated with stars. The nebular hypothesis had already established that stars would condense from star clouds. What happened after that, though, required knowledge of how stars generate their energy.

The consequence of Hans Bethe's work was that astronomers realized what would happen as a cloud of gas collapsed. It starts as a large cloud like the Orion Nebula (see figure 1.2), which is 26 light years across and 1,500 light years away and has a mass some thousands of times the mass of the Sun. In the center of this cloud, the density might be as high as 600 molecules per cubic centimeter. Although this is still a very high vacuum, it is about 1,000 times greater than the density of interstellar space. Initially this gas would be extremely cold but, as it collapsed, it would start to warm and get much denser in the center (see figure 1.3). When the temperature has reached about 10 million degrees centigrade and the density of the *gas* in the center of the cloud is around six grams per cubic centimeter (six times the density of water), nuclear fusion will start in the p-p chain, although the deuterium reaction will start at much lower temperatures, around 50,000°C.[13] At this point the star will suddenly switch on.

When the star switches on, various things happen. The temperature of the nucleus will rise sharply to around 14 million degrees and will start generating enormous quantities of energy. The Sun converts some 500 million tonnes of hydrogen into helium each second, losing as it does so 4 million tonnes of mass per second. This is converted into the energy that the Sun liberates.[14] This energy in turn generates a huge pressure on the falling gas and stops the cloud from collapsing. The gas cloud thus becomes a stable star, with the huge pressure exerted by the radiation from the core exactly balancing the mass of the gas and the pull of gravity from it that makes it collapse.

Because 500 million tonnes of hydrogen per second sounds like an enormous mass to lose, one might wonder if the Sun is going to run out of hydrogen quickly. The mass of the Sun is so huge, though, that the supply of hydrogen in the center will last for some 10,000 million years—twice the current age of the Earth. Thus, far from being a very old star as the collapse theory had stated, the Sun is middle-aged and in the prime of its life. If we compare the Sun to an average human, you might say that it has an age equivalent to a 40-year-old who has looked after his health well and is in a stable and satisfactory life situation.

We know what happens *when* a star collapses, but we have said nothing about *why* it collapses. Why should a large and apparently stable gas cloud suddenly collapse anyway? Different ideas have been proposed over the years, but there are two ways of making a cloud collapse that have been observed in space.

The first idea is one that was proposed to answer a critical question about stars. A cloud like the Orion Nebula is mainly made of hydrogen and helium gas, yet Earth and the other planets contain huge amounts of other, much heavier elements. Where do these come from? We know that heavy elements are formed in the interior of the largest stars and spread through the interstellar medium when these stars explode as supernovae. What happens, though, when a supernova happens to be close to or within a nebula such as M42? The answer is that a shock wave and a cloud of dust and gas moving at perhaps 10 percent of the speed of light will slam into the nebula. The material of the nebula will be swept up by this blast of material and will mix with it, becoming seeded with huge quantities of heavy elements in the process.

An idea of how this process works can be seen in figure 1.4. It shows the expanding shell of dust and gas from a supernova that exploded in Cygnus perhaps 5,000 years ago. The shell is still slamming into the interstellar gas and compressing it, glowing blue as it does so. When the shock wave runs into a nebula as dense as the Orion Nebula, the impact is considerable and causes an intense compression at the point of impact. This impact destabilizes the cloud and starts its gravitational collapse. Once collapse has started, the formation of a new star or stars is almost inevitable. Sometimes at least, the death of old stars gives rise to the formation of new ones.

Considerable evidence suggests that the formation of the Sun was triggered by a nearby supernova some 5,000 million years ago and that the supernova had enormous long-term effects. Quite apart from seeding the protostellar cloud that gave rise to the Sun and our solar system with iron, silicon, magnesium, sodium, and the like, the supernova also seeded the cloud with highly unstable radioactive elements. Of radioactive elements, one of the most important is aluminum-26. Normal, stable aluminum, known as aluminum-27, has 13 protons and 14 neutrons. In the storm of radiation from a supernova explosion, however, huge quantities of aluminum with only 13 neutrons, aluminum-26, is also formed. This alternative form of aluminum, known as an isotope, is extremely unstable and disintegrates into magnesium-26 (12 protons and 14 neutrons) in a few tens of millions of years. In nature, though, it is the lighter isotope of magnesium, magnesium-24 (12 protons and 12 neutrons), that is the most common. However, if we look at the oldest material in the solar system, the globules of rock called chondrules found inside some meteorites,[15] we discover that they have far more of the heavy isotope magnesium-26 than is normal in nature. This indicates that something in the distant past caused large amounts of aluminum-26 to come into existence at the time that the solar system was starting to form. Because aluminum-26 is so unstable, it has to have formed at the same time that the cloud started to collapse and in the same place; had it come from a more distant supernova, all the aluminum-26 would have decayed to magnesium long before it came to be incorporated. So the evidence suggests strongly to astronomers that a supernova triggered the formation of the solar system.[16]

In recent years, a second idea has become popular again. In figure 1.5 we see what has become one of the most famous images ever taken by the Hubble Space Telescope. It shows the central region of the Eagle Nebula—Messier 16 in Serpens—which is a nebula similar to the Orion Nebula. The stellar wind from giant, very luminous young stars off the top of the image has eroded the nebula, pushing most of the gas away. Three narrow columns remain, though, where, in the peak of the column, there is a much denser cloud of gas inside that protects the region below from being eroded away. Each of these dense clouds is a star or a small cluster of stars in the process of formation. In this case the factor that causes the collapse of the cloud is the impact of the intense stellar wind.

Stars usually form in groups. The reason for this is simple. In most cases, the region of the nebula that collapses is relatively large and may have a mass tens, hundreds, or even thousands of times the mass of the Sun. As it collapses, random effects will ensure that the gas will almost always start to form a number of different vortices where it starts to concentrate. As the gas concentrates in each vortex, gravity takes over, and the vortex will accumulate mass, each one being the seed for a new star. The greater the concentration of mass, the greater its ability to attract more mass; thus a larger seed will inevitably form a much more massive star. We know, however, that less massive stars are much more common than more massive ones, but they are less easy to see because they are so faint. If we were to plot all the stars in our Galaxy together in a giant H-R diagram such as that in figure 1.1, the bottom right hand corner would contain almost all of them. There are a million times more stars one-tenth of the mass of the Sun than there are stars 100 times the Sun's mass. There are 2,000 times more stars that are the same mass as the Sun than there are stars that are 10 times the Sun's mass. Astronomers know of 64 stars, including our Sun, within 5 parsecs (16.3 light years of the Earth). Of these, just 3 are more massive and brilliant than the Sun, and no fewer than 39 are tiny, dim red dwarf stars with only a small fraction of the Sun's mass. This happens because smaller seeds are "starved" by nearby bigger ones. We see this in our solar system where the formation of massive Jupiter stopped the formation of anything larger than the small asteroids in the next orbit in from the Sun.

Although few in number, these massive and luminous stars have a profound effect on the gas cloud. Their intense radiation and powerful stellar wind pushes away the remaining gas and dust around them, opening a hole in the cloud. Stars form in the middle of the cloud and progressively push away the gas, opening a hole or bubble in the nebula, inside of which we can see the stars clearly. Over millions of years almost all the gas and dust is used up to form stars, or blown away, until after a few hundred million years no trace of the original cloud remains (figure 1.6). Over time, a cluster of stars like this will break up as individual stars escape the pull of gravity of the rest of the cluster and slowly spread out into space until all trace of the original cluster has disappeared.

So, when stars form, it is as a group or cluster of perhaps hundreds or even thousands of stars, but usually this group has a few large stars that dominate a majority of much smaller and less prominent stars. It is salutary to remember that in the Pleiades Cluster, the famous Seven Sisters in Taurus that can be seen easily with the naked eye, there are several hundred much smaller and dimmer stars apart from the six brightest stars that anyone with normal eyesight can see.[17] In fact, the difference in brightness between the brightest and the faintest stars in the cluster is at least a factor of a million.

(Most) Old Stars Never Die, They Simply Fade Away

Stars are born with a huge range of sizes, from the tiniest, which are less than a tenth of the mass of the Sun and a millionth of its luminosity, to those veritable giants 100 times more massive than the Sun and 100,000 times more luminous. How does this affect their deaths? We have all heard of the dangers of obesity and the risk of premature death that it causes—in particular, that obesity is a major cause of heart disease. What many people will not realize is that the stars themselves are not immune from these problems. In fact, obesity and heart disease are a major cause of death in the stellar community.

How a star dies depends on two principal factors: its size and its family situation.

The overwhelming majority of stars die quietly and fade slowly away. Not for them is there a final blaze of glory. However, the largest stars have a violent and spectacular end, particularly if they have an unstable family situation.

Let us look first at an "average" star, by which we actually mean 99.9 percent of all stars in the Galaxy.[18] A star like the Sun will continue to convert hydrogen to helium in its center for 10,000 million years—at present it is about half that age—until approximately 7 percent of all the hydrogen has been converted to helium. This helium is inert and accumulates in the core of the star like ash in a fire. When this critical point is reached, the helium ash will choke the hydrogen reaction. Until then, the force of gravity is in a constant stalemate against the pressure of the superheated gas in the core of the star: gravity tries to make the star contract, while the pressure of the gas tries to make it expand. When the nuclear fusion reaction stops, gravity suddenly encounters much less resistance, and the core of the star contracts suddenly and violently. As it contracts, its temperature races up until a new equilibrium is reached; for stars the size of the Sun, this comes when the temperature is so high that the helium starts to combine to produce carbon in a "triple-alpha" reaction, in which three helium nuclei come together at once to form a carbon nucleus. This gives the star a new source of energy.

When this happens a curious change will come over a star. Its interior will be very much hotter than before, but the force of gravity remains the same. More heat and pressure than before, though, will be available to inflate the star. The result is that the outer layers will be blown outward. The star swells to an enormous size and turns into what is termed a red giant, with a small and dense core surrounded by an immensely swollen and much cooler shell. In the Sun's case, when it becomes a red giant, it will grow until Mercury, Venus, and Earth are swallowed up inside. Over a few tens of thousands of years, the outer layers of the Sun will be shed progressively and lost into space, and the Sun may lose a significant proportion of its mass.

For smaller stars, the crisis takes longer to arrive. We might think that because a less massive star has a much smaller amount of hydrogen, it must use its fuel up more rapidly. The surprising thing, though, is that the smaller the star, the slower the hydrogen reaction runs and so the longer its limited supply of hydrogen will last. The smallest red dwarf stars will continue to "burn" hydrogen for at least 100,000 million years. In contrast, the larger the star, the more rapidly it uses its hydrogen, and the less time it will last.

The Sun will not enjoy its new stability using helium as fuel for long. After a few 100 million years, the carbon that builds up in its core will choke the helium reaction. This time there will be no reprieve from the force of gravity. The smallest stars will never even get hot enough in their cores to initiate helium reactions and so will not earn even this brief respite. A star a few times the mass of the Sun may go through a further "collapse and restart" phase, briefly converting carbon into nitrogen, oxygen, neon, magnesium, and even silicon before finally running out of usable fuel. Whatever the star, there will come a point where there is no more usable nuclear fuel, and death will follow when the core collapses as gravity finally wins the battle. The reason why the collapse stops in the end is the atomic force called *degeneracy*. This is produced when atoms are forced so close together that the orbiting electrons are pushed almost back into the atomic nucleus. The result is a white dwarf star. This is the dead shell, no bigger than a large planet, of a star that continues to glow from the residual heat that it contains. A white dwarf will cool slowly until it disappears completely. If the star has been large enough to convert helium into carbon, its nucleus may even crystallize over millions of years into an immense diamond, although one that no earthly prospector can ever reach, buried deep in the heart of a cold, dark star.

What if the star is much more massive than the Sun? In this case, its demise can be very much more spectacular. What happens to a star like Betelgeuse in Orion, which is more than 20 times as massive as the Sun? Such a massive star will use its hydrogen in a few tens of millions of years because, although it has more fuel than the Sun, it uses it so much faster that it will last a fraction of the time. Betelgeuse has passed through the helium crisis and also the carbon crisis. In fact, in its core there is a nucleus of silicon at a temperature of perhaps 1,000 million degrees. Betelgeuse is in the final stage of nuclear reactions where silicon is being converted into iron in its core. Around this core—where there are still other, lighter elements—there is a layer where carbon is still being converted to oxygen, neon, magnesium, and silicon. Farther out still, there is a layer where helium is being converted to carbon, and, finally, outside all these, is a layer where hydrogen is being converted to helium. All these layers like onion skins are surrounded by a huge cloud of hydrogen, which is too cool and thin to react.

Iron is the most stable of all the elements; any nuclear reaction involving iron uses energy instead of producing it. So, when the core of the star has filled with iron, it reaches a final and definitive crisis. Once more the star will collapse as gravity takes over. In perhaps as little as a second, the star will implode. However, this time two things will happen that have not happened before. First, nothing will stop all the mass of the star crashing at enormous velocity into the center in a train wreck of phenomenal magnitude. Second, when all this material reaches the center, suddenly there is a huge quantity of unreacted elements—hydrogen, helium, carbon, and the like—in a tiny space at unimaginable pressure and temperature. All this gas crashes into the center and rebounds as if it has hit a solid brick wall, aided by an immense orgy of nuclear reactions. Some 90 percent or more of the mass of the star is blown off into space at velocities that may be 10 to 20 percent of the speed of light in the resultant explosion, which we know of as a supernova or, more exactly, a type II supernova.

For a short time a supernova may produce as much energy as an entire galaxy of 100,000 million stars. In its center remains what is left of the core of the star. If this is less than three times the mass of the Sun, it will be converted into a neutron star. The pressure of the supernova explosion is so immense that all the protons and electrons in the nucleus combine to form neutrons. A neutron star has all the mass of a star compressed into a diameter of perhaps 10 kilometers. Surrounding the neutron interior is a thin crust of iron and, above that, a dense atmosphere just a few centimeters thick! Neutron stars spin rapidly and, if they are aligned toward the Earth, we may see them flashing once every rotation as the magnetic pole passes in front. Usually this is detected as a radio pulse, and the star is known as a *pulsar*. One pulsar that flashes three times a second was even likened by its discoverers to a cosmic Ringo Starr, beating out time to some unheard celestial music.

There still remains one more spectacular possibility. If a star is large enough—and Betelgeuse is close to this point—the remnant left after the supernova explosion may be several times the mass of the Sun. A neutron star cannot have a mass greater than three times the mass of the Sun because, if it does, the nuclear forces that keep the neutrons apart will be weaker than the force of gravity. Should the remnant exceed three times the mass of the Sun, we believe that no force of nature can prevent it collapsing into a black hole. Thus the fate of the very largest stars on dying may be literally to disappear from the Universe as black holes.

What about stars in families? Many stars are double or binary, with two stars rotating around each other. What happens when such stars get old? In this case we may see a different kind of supernova that is visually brighter but has less substance to it and is an entirely less violent explosion than the explosion of a massive star—one might say that it is all flash and no substance.

Where two stars are of different mass, the larger one will use up its nuclear fuel first and turn into a red giant. If the two stars are close enough together, the smaller of the two will start to feed off the red giant star, pulling material off it onto itself. Astronomers call this process *mass transfer*. Suppose the larger star is four times as massive as the Sun and the smaller twice the Sun's mass. After this process has ended, the larger star may have shrunk until it is just the same mass as the Sun, while the other has swollen to be five times the Sun's mass. The now-shrunken star will collapse into a white dwarf, while its companion will enjoy a suddenly greatly accelerated life, becoming, in its turn, a red giant star. Now the process of mass transfer will reverse, and material will fall from the red giant onto the white dwarf, increasing its mass once again. If the mass of material falling onto the white dwarf is too large, however, it will suddenly become unstable. Degeneracy pressure will only support a star up to 1.4 times as massive as the Sun—this mass is called the Chandrasekhar limit, named after the Indian astrophysicist, Subramanyon Chandrasekhar. If the mass increases beyond this limit, the white dwarf will collapse in an immense nuclear deflagration. All the hydrogen that has fallen on it will combine into helium, leading to a supernova explosion, with a neutron star and a huge expanding gas cloud as the result. This is a supernova of type Ia.[19] Visually it is much more spectacular than a type II supernova, as almost all the energy goes into the visible flash. In a type II supernova, much of the energy is produced in the form of neutrinos or an expanding dust cloud. What is especially interesting about type Ia supernovae is that almost all seem to have just about exactly the same intrinsic luminosity, which is logical because the Chandrasekhar limit should be the same for all stars. This makes them a valuable tool for astronomers measuring the distance to remote galaxies, for we only have to measure the supernova's brightness to calculate how far away it must be.

THE BIRTH AND DEATH OF STARS is a fascinating subject. Although we now know what powers the stars and makes them shine, there are many aspects of their birth and death that we do not properly understand. Ideas about both stellar birth and supernovae have changed radically in the past 30 years and are continuing to change as astronomers find new ways to test their theories. It could be that in 30 years time much of what is written here will become outdated. Even now a new theory is rising up that challenges the established one for type Ia supernovae. It suggests they may be due to the collision of two white dwarf stars—what is termed the *double degeneracy model*. It is now time, though, to pass on to even more exotic objects and look at where they come from and what they may be.

CHAPTER 2

How Do We Know That Black Holes Exist?

Astronomers talk about many exotic objects in the Universe: neutron stars, pulsars, quasars, blazars, and gamma ray bursters, to name a few. None, though, excites public interest quite as much as black holes. A black hole, though, has its own special problems: nothing can escape from them, not even light, so they are intrinsically invisible. How can we know that an invisible object like a black hole even exists? How can we speculate what happens inside a black hole?

Black holes have strange effects on time and space. Some scientists speculate they could be gateways to crossing the Universe at velocities far greater than that of light—they may literally be shortcuts across space. Science fiction has used black holes and their exotic properties a great deal. In the 1980s the Disney studios introduced a whole generation of youngsters to them with their film *The Black Hole*. Joe Haldeman had his soldiers of the future hopping from battle to battle in his novel *The Forever War*, using black holes as a transport system. In the novel and film *Contact*, Carl Sagan's heroine is transported across 25,000 light years of space instantly to the center of the Galaxy using a futuristic metro that links different points of the Galaxy through massive black holes.

Although the idea of a black hole was first proposed in the eighteenth century, it is only in the past 30 years that astronomers have really begun to take them seriously. Since then, black holes have become one of the most common and accepted explanations for a whole range of phenomena. Some astronomers started to question whether the answer "black hole" is a knee-jerk reaction whenever we are asked to explain any unusual phenomenon. There has been resistance by a portion of the scientific community to accept that black holes could really exist. Instead, they proposed all manner of alternative explanations for unusual objects that were observed. Most scientists by now, though, have accepted the existence of black holes in the Universe, but there are still some holdouts who regard their existence as unproved.

Just what evidence do we have that black holes really exist, and how can we possibly observe something that, by definition, we cannot see?

Thinking of Black Holes

Although the term *black hole* is relatively recent, the idea of black holes actually dates back more than two centuries. As early as 1796 the French mathematician Pierre Simon Laplace was studying the subject of the escape velocity—that is, the minimum velocity at which a body has to be launched to escape completely from an object and never return. If you go outside into the yard and throw a ball in the air, it will rise to a certain height and then fall back; the Earth's gravity pulls on it and slows the ball until it stops and falls back. The harder that you throw the ball, that is, the faster you throw it upward, the higher it goes before gravity finally brings it to a halt and drags it back. If, instead of throwing the ball, you use a cannon or a mortar to launch it, the ball will go even higher before falling back. Laplace worked out that if a ball could be thrown at just over 11 kilometers per second, it would never fall back to Earth.[1] This is our "escape velocity." Though some people say, mistakenly, that the object has escaped from the Earth's gravity, it has not. Earth's gravity is still there and slowing it down; however, it will never manage to stop and pull back an object that has left at more than 11 kilometers per second. Launch a rocket at that speed, and it will never return.

Laplace took things further. He realized that the escape velocity depended on the *size* and the *mass* of the object. Make the Earth a quarter of the diameter but maintain the same mass, and its escape velocity would not be 11 but 22 kilometers per second. Suppose you kept squeezing the Earth, making it smaller and smaller. Finally you would reach a point, with the diameter reduced to 1.8 millimeters, at which its escape velocity would reach the velocity of light. Laplace knew that at this point the Earth would disappear because light itself would no longer be able to escape from it and allow it to be seen. This is what we know of as a black hole—an object that has such an enormous gravity field that light itself cannot escape from it. A more massive object such as the Sun would be larger when it became a black hole—you would have to squeeze the Sun's 1,392,000 kilometers down to 2.95 kilometers to raise its escape velocity to the velocity of light.

Laplace did not know that the velocity of light is a physical limit that no object can exceed—in other words, that a property of black holes is that nothing at all can escape from them. This physical limit was established by Albert Einstein in 1905 with his special theory of relativity.[2]

After years of further work, in 1915 Einstein published his general theory of relativity, which deals with gravitation. This work came to the attention of a German mathematician and astronomer named Karl Schwarzschild. In 1909, though still in his 30s, Schwarzschild had become director of Potsdam Observatory in Germany but left his post during the First World War, when, because of his relatively young age, he was drafted into the German army. While serving on the Russian front, he died in 1916 at the age of 43. One of the key conclusions of the general theory of relativity was that a ray of light would be bent by a gravitational field. Schwarzschild took this result and, while at the front, worked on two papers that developed the concept. Ignoring the horrors of the war around him,[3] he worked out that there was a certain radius of object for any given mass at which light would be unable to escape. This radius is now called the Schwarzschild radius in his honor and is defined as

$$R_s = 2GM/c^2,$$

where G is the universal constant of gravitation, M is the mass of the object, and c is the speed of light.

Schwarzschild continued looking at the characteristics of objects smaller than the Schwarzschild radius. In his general theory of relativity, Einstein had proposed a series of equations, which we now know as the Einstein field equations,[4] to describe the gravitational field produced by any mass in space. These equations were so complicated that, when he published them, Einstein was convinced they would never be solved. He could little have imagined that within a year another scientist would prove him wrong and, even less, that it would be a scientist who, far from the peace of a university or an academic hall, could work on the equations only in the lulls between dodging Russian shells and bullets. Schwarzschild solved the equations by making two simple approximations: he assumed, first, that the black hole is not rotating and, second, that it has no electrical charge.

This solution became known as the Schwarzschild black hole. The case is not only the simplest possible but is also strictly theoretical because the object that collapsed to form the black hole would have to be totally stationary with no rotation whatsoever; this does not happen in the real world as all the stars, planets, and galaxies that we know of are rotating. We also know that, just like the skater in a spin who pulls in his or her arms, any object that we make smaller spins much faster. (Astronomically, the best demonstration of this involves neutron stars, which are the remnants of stars destroyed in supernovae; even though the star that exploded was rotating very slowly, by the time that it has been compressed into a neutron star it is rotating at breakneck speed, often hundreds of times per second.) It is hard to think of how to form a black hole that is not rotating.

The less simple the black hole, the more complicated the equations. Once rotation and electrical charge are added to a black hole, solving the equations becomes a far more difficult proposition. In 1916 a German physicist Hans Reissner proposed a solution for an electrically charged black hole. This solution was refined by the Finn Gunnar Nordström in 1918, and so these are now known as Reissner-Nordström black holes.

Solving these same field equations for a rotating black hole took many years, but in 1963, the New Zealander Roy Kerr found a solution, and a rotating black hole is now known as a Kerr black hole in his honor. Within two years of Kerr's work being published, even the most complicated case of a charged, rotating black hole was solved by a team of scientists led by the American Ezra Newman, and these objects are now known as Kerr-Newman black holes.

So, black holes were firmly established as a theoretical possibility by scientists before the end of the First World War. For many years, they seemed to be just that —a theoretical possibility. Astronomers little suspected that only a half century later black holes would become a reality, used to explain many unusual astronomical phenomena, although a small group of astronomers led an increasingly desperate rearguard action to explain objects in ways that did not require a black hole.

The Violent Universe

Until the 1960s astronomers had no need or use for black holes. The Universe seemed to be a remarkably simple place filled with stars organized into galaxies and probably untold millions of planets around those stars; there were no strange and exotic objects. Even the newly discovered radio stars seemed likely to be normal, and when objects like Cygnus A were identified by astronomers, they looked like simple colliding galaxies.

The first hint of the violent Universe around us unfolded in 1962. Back in September 1949 a team from the Naval Research Laboratory, led by Herbert Friedman, had mounted Geiger counters in a captured V2 rocket that was fired above the Earth's atmosphere. The Geiger counters registered weak x-ray radiation from the Sun's corona, where the exceptional temperatures of around 1 million degrees heat the gas to such a high temperature that it shines in x-rays. X-ray astronomy was born. The 1962 experiment was more ambitious; a team led by Richard Giacconi at American Science and Engineering in Cambridge, Massachusetts, used a small x-ray detector aboard an Aerobee rocket. Its aim was to look for x-rays from the Moon. Giacconi, now at Johns Hopkins, and his colleagues knew that the Moon was constantly bathed in cosmic radiation and that the impact of this high-energy radiation could cause the lunar rocks to fluoresce in x-rays, thus revealing its composition.[5] Chance took a hand, with astonishing consequences. Because the scientists were uncertain whether or not they could point the rocket correctly, it was decided to roll it so that it would scan the sky, crossing the Moon during the roll. The scientists did not detect emission from the Moon because it was too weak. To their amazement, though, not only did they detect a faint glow of x-rays from all round the sky, but also from a strong source of x-ray emission in the constellation of Scorpius. Aping radio convention, this source became known as Scorpius X-1, or Sco X-1—the first x-ray source detected in the constellation of the Scorpion.

This discovery was important because generating x-rays requires tremendous energy—extremely hot gas. The solar corona is heated to temperatures high enough to produce x-rays because it is extremely tenuous and thus requires relatively little energy to heat it so much. Move the Sun even a little farther away, though, and its x-ray emission would be too faint to detect. For a source that was as many light years away as Sco X-1 was, to have the quantity of x-ray emission that was detected showed that it must generate huge quantities of x-rays. If produced in the simplest manner, by matter getting heated by falling onto an object and being accelerated by gravity, the object had to be exceptionally small and dense.

Whether we were ready or not, astronomy was about to get interesting. High-energy astrophysics had arrived.

Further flights showed that there were several other sources of x-rays in the sky: the Crab Nebula was one; so too were the radio galaxies Centaurus A and Virgo A. The Crab Nebula was, for many years, used by x-ray astronomers as their reference object, and they talked of the brightness of other sources in terms of what fraction of the brightness of the Crab Nebula they are—a source had a brightness of "0.5 Crab," or "0.1 Crab," or "2 Crab."

Through the 1960s, discovery after discovery showed astronomers that our Universe is not only stranger than we imagine but stranger than we *can* imagine— and unimaginably more violent than anyone had believed. The first of several shocks was the discovery of quasars. These looked just like a star but are thousands of millions of light years away, and although they were measured to be no more than a few tens of light years across, they can emit as much energy as 100 galaxies. This was distinctly against the rules—how could something just one-ten-thousandth the diameter of a galaxy emit as much energy as 100 galaxies of stars? Astronomers played with all kinds of theories, from superstars to antimatter annihilation, to explain the source of the energy. By 1964 the American astrophysicist Edwin Salpeter had already suggested that the source of the prodigious energy of quasars might be matter falling onto a black hole, although the words "black hole," then rarely in scientific usage, did not appear in the text[6] (Salpeter claims, with justification, that he was the first to suggest this possibility and, 40 years later, although well cited, his pioneering article is less well known than it should be). The time was not yet ripe for scientists to take the concept seriously— it would take another five years for that to happen.

What happened to make such revolutionary ideas as black holes not just respectable but *necessary* were two things. As so often in life, one remarkable, but not too unbelievably exotic a discovery, made another far more outrageous theory much more plausible.

The first discovery was that of pulsars. In September 1967 a young graduate student at Cambridge University, Jocelyn Bell, became aware that a little patch of unusual noise or "scruff" kept reappearing in the 30 meters of paper chart that the radio telescope she was controlling produced every single day. The noise seemed to come from the same part of the sky each time it appeared in the paper trace. By October, it seemed time to investigate the mysterious object further and make a high-speed recording of its signals. Unfortunately, some mischievous demon intervened, and the mystery source disappeared for several weeks. Finally, at the end of November it returned, and she got the fast recording. She found, to her considerable surprise, that the source was emitting a series of pulses every one and one-third seconds. Here, once again, fate took over. Jocelyn Bell's Ph.D. supervisor, Anthony Hewish, thought that the signals had to be man-made and caused by some kind of interference; Jocelyn Bell, in her self-confessed ignorance, thought that they might be signals from an unusual star. Bell was right but scarcely imagined how unusual the star might be. In December Bell noticed a second similar object in a completely different part of the sky. This eliminated any faint suspicion that the mystery signals might be from an extraterrestrial intelligence. Then in one single day after Christmas, two more appeared on the same piece of paper.

Anthony Hewish gave a seminar at Cambridge about the strange new objects shortly before the discovery was announced formally by its publication in the journal *Nature*. Immediately Fred Hoyle suggested at the end of the seminar that he thought they were produced by supernova explosions. This was a brilliant piece of instant analysis and was rapidly confirmed by other researchers. Within weeks astronomers were certain that the newly named *pulsars* were neutron stars—that is, stars that are the remnant of the core of a giant star that has exploded as a supernova, crushing its center so violently that all the atoms have been compressed until the protons and electrons combine to make neutrons. A neutron star has all the mass of a star as much as three times more massive than the Sun, crushed into a diameter that may be only about 10 kilometers and, to boot, rotating extremely rapidly as the star's spin accelerates to match its shrinkage. The pulsar sends a beam of radio emission from its poles like a lighthouse beam, and each time the beam passes over us we see a flash.

The discovery of pulsars showed that compact objects, far beyond white dwarf stars, did exist. From there it was a simple step to think of yet-more-compact objects.

The second crucial moment was the launch of the x-ray satellite Uhuru on December 12, 1970. It was launched on Kenyan independence day from the San Marcos launch platform off the coast of Kenya (hence, the name *uhuru*, the Swahili word for "freedom"). Uhuru had the task of mapping the x-ray sky, searching for objects that emitted x-rays. The satellite found 339 sources that were at least one-thousandth as bright as the Crab Nebula. Some were galaxies and groups of galaxies, or supernova remnants, but most of them were x-ray binaries. In these, ordinary stars orbit neutron stars that emit x-rays as material falls from the ordinary star onto the neutron star and gets extremely hot in the process. However, one of the Uhuru sources, Cygnus X-1, did not appear to fit this profile.

When there is a binary star, the signal you detect changes periodically as one star goes around the other or, more correctly, around the center of mass of the two. By measuring this period we know how long it takes to make an orbit around the center of mass. Astronomers can then calculate the total mass of the two stars using Newton's laws of gravitation. In these systems one of the two stars—the normal one—will be visible and observable with a normal telescope, while the other—the compact companion—is invisible. If we can observe the visible star and see what type it is and how luminous it is, we can estimate its mass accurately and so infer the mass of the compact star.

In most cases it was evident that the invisible object was less than three times the mass of the Sun, which is the limit to be a neutron star (a more massive object would collapse into a black hole because the neutrons themselves are not strong enough to support the mass of the star), and thus was a neutron star. In Cygnus X-1, though, the invisible companion has a mass that was initially estimated to be five to eight times the mass of the Sun. Such a huge object is far too large to be a neutron star. A minority of astronomers suggested it might possibly be a pair of neutron stars,[7] but most accepted that it was almost certainly a black hole. Over the years estimates of the mass of the invisible companion improved. Our current estimate is that it is at least 10 solar masses (some estimates make it as much as 16 times the mass of the Sun). By no stretch of the imagination can it be anything other than a black hole.

Uhuru observed x-rays from the sky that were of relatively low energy as x-rays go. Astronomers measure the energy of these photons—the individual particles of light—using units of electron volts (1eV is equivalent to 1.6×10^{-19} joules of energy, and 1 joule is the kinetic energy of a 2-kilogram mass moving at 1 meter per second). The higher the number, the more energetic the photon. The x-rays observed by Uhuru went from 2 to 20 KeV—that is, thousands of eV. Satellites were launched capable of detecting higher and higher energies—first hundreds of KeV; then on to gamma rays, which register millions of electron volts (MeV); and on to high-energy gamma rays, which have an energy of hundreds of MeV.[8] This was the astronomical equivalent of progressing from a bow and arrow to bullets and then on to heavy cannon.

As astronomers passed from low-energy x-rays—what they call "soft" x-rays—to high-energy x-rays, or "hard" x-rays, they found the number of sources in the sky reduced sharply. This was not just because the hard x-rays are more difficult to detect, but because they emit so few x-ray photons.[9] There are many x-ray binaries that emit soft x-rays. There are far fewer that emit large quantities of hard x-rays. And by the time we get to the hardest x-rays and gamma rays, there are very few. Which sources will emit the highest energy x-rays? Evidently those where there is the most massive companion star that accelerates the material falling onto it most. In other words, the distribution of these x-ray binaries suggests to us that most contain an invisible star of quite low mass, but a few have far more massive objects—almost certainly black holes—in them.

So, How *Do* You Pin Down a Black Hole?

On May 22, 1989, the Japanese x-ray satellite Ginga detected an unusual new source of x-rays in the constellation of Cygnus and reported it as an x-ray nova. This source was given the catalog name GS2023+338 and astonished astronomers by showing huge variations in brightness over just a few minutes. Soon it was recognized that the source had also brightened in visible light and that it was identical with a star that had appeared some 50 years earlier. This had reached magnitude 12.5 and was recorded as a nova, having been cataloged as Nova Cygni 1938 and then renamed V404 Cygni.[10] Coincidentally, the 1989 outburst also reached magnitude 12.5. Astronomers were in for some very interesting times as telescopes all around the world started to observe the strange new star.

Astronomers were initially disconcerted and assumed that it was a low-mass x-ray binary of a previously unknown type, but the spectrum of its light was unlike anything ever seen before. As studies continued, it became obvious that something curious was going on. In 1991 Phil Charles, Tim Naylor, and my ex-colleague

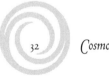

Cosmological Enigmas

Jorge Casares used the 4.2-meter William Herschel Telescope in La Palma to study V404 Cygni. Their results were fascinating. They used the telescope's spectrograph to measure the Doppler shift of the star, that is, how much the light of the star is red-shifted, or moved to longer wavelengths, as it moves away from us in its orbit and how much it blue-shifted, or moved to shorter wavelengths, as it swings toward the Earth (this is exactly the same effect as hearing the tone of a siren rise and fall as a police car races past). They found that it varied by an amount equivalent to a velocity of 211 kilometers per second every 6.473 days. This meant that both the stars' orbital period and their speed were known, so the masses of the two stars could be calculated with some precision. They found that the invisible companion star could not be smaller than 6.3 solar masses and was probably between 8 and 16 times the mass of the Sun, even if the two stars were orbiting each other exactly in our line of sight so that the Doppler shifts were the biggest possible. (If the orbit of the stars is inclined, we see the Doppler shift only from the part of the movement that is directly toward us, so the real velocity of the stars in the orbit would be much faster.) Later they were able to estimate the inclination of the system and calculate a better estimate for the mass for the invisible star of 11 times the mass of the Sun. It could only be a black hole.

In additional studies to estimate the masses of the two stars in the system, the same group has concluded that the invisible black hole must be at least 25 times the mass of the visible star, which itself is an orange subgiant star a little more massive than the Sun. Estimating that the black hole may be as large as 28 times the mass of the Sun, the group has concluded that it represents an "extraordinary system."

V404 Cygni thus received the honor of being the first x-ray binary in which the invisible companion was proved to be a black hole. No other theory has been proposed that explains what an invisible object of more than 6 times the Sun's mass, let alone 28 times, could be if it is not a black hole. So, given that V404 Cygni must be a black hole, how did it get there?

The answer is simple. The binary must have started as two massive stars, one of which evolved extremely rapidly and turned into a supernova, blasting 80 to 90 percent of its mass into space and quite probably at least part of its companion star. During the supernova explosion, the core imploded with such violence that it was converted into a black hole. For this to happen it must certainly have been greater than 20 times the mass of the Sun, perhaps very much greater, because a smaller star would have left only a neutron star behind. The black hole was initially quite small, but as the two stars were close together, material from the smaller star trickled onto the black hole. This trickling made the black hole increase in mass, while it drained mass from the secondary star, which proceeded to shrink. Right now we are close to the end game—the secondary star has shrunk so far that it has very little more mass left to lose, although mass is still streaming constantly onto the black hole's accretion disk, which is what causes the massive flare-ups of x-rays.

V404 Cygni is not the only black hole in our Galaxy. There are 16 other x-ray systems in which the invisible star is at least six times the mass of the Sun and so is probably (though not certainly) a black hole. Figure 2.1 shows an artist's concept of what one such system may look like, alongside the view that astronomers get through an x-ray telescope.

If black holes are so common, are we in danger from one? The answer is most definitely no. The closest black hole candidate is Nova Monocerotis 1975, also known by the catalog name A0620-00; this is at an estimated distance of 2,700 light years. Its event horizon is 200 kilometers across, so we would have to get very much closer to be in any danger. But what about unknown black holes? Could there be an undetected black hole nearby that is a threat to us?

To notice the effects of a black hole, one has to be extremely close by. From a distance as little as 1,000 times the Schwarzschild radius of the black hole, there are no obvious effects. Luckily, while black holes have a very strong field of gravity, they also have an extremely large *gradient of gravity*—get just a little further away from it, and the effects of its gravitational pull reduce enormously; if you cannot see its effects, you are absolutely safe from it.

Bigger and Better Black Holes

Even before there was definitive proof of the existence of black holes in our own Galaxy, astronomers speculated about the existence of giant black holes in other galaxies. In 1969 the British astronomer Donald Lynden-Bell wrote an article for the journal *Nature* in which he really set the bases of the theory that quasars were galaxies with a giant black hole in the center. This central black hole was swallowing material from the galaxy and producing enormous quantities of energy in the process. Although Salpeter had proposed a similar idea in 1964, it was

Lynden-Bell's paper that made the idea respectable. He suggested that the black holes responsible might be millions, or even hundreds of millions, of times the mass of the Sun.

Lynden-Bell's idea was widely accepted, despite some resistance, because, with the discovery of neutron stars and compact x-ray binaries, astronomers were now ready to believe that even more exotic objects could exist. Lynden-Bell suggested that there might be black holes of millions or even hundreds of millions of times the Sun's mass hidden in the centers of galaxies. How, though, to prove this?

Apart from the prodigious energy of quasars, which seemed to demand that exotic objects were responsible, several other things made astronomers suspect that galaxies hid giant black holes. One was the way that galaxies rotated. Astronomers could measure the speed of rotation of galaxies at different distances from the center. In the Andromeda Galaxy, for example, they could see that stars in the center were rotating rapidly around the center and could work out how massive the center of the galaxy had to be to allow them to move that fast. They could also work out from the brightness of the center of the galaxy how many stars there were there. In every case, it was found that the stars rotated far more rapidly than could be explained by the mass of visible stars. These galaxies had some huge dark mass in the center that could not be observed, one that had a strong force of gravity. The obvious suspect was a black hole lurking unseen in these galaxies.

Most convincing of all, though, were the discoveries that radio astronomers were making about different distant active galaxies. Many radio sources featured a tiny, concentrated region—no more than a few light years across—that generated a large part of the energy and then one or two straight jets in opposite directions pointing on opposite sides of the tiny central source. A typical case is the radio galaxy Virgo A, also known as M87 as it was cataloged by Charles Messier in the eighteenth century. Radio observations had shown a jet coming out of the center of the galaxy, aligned very accurately with the radio source in the exact center of the galaxy. The Hubble Space Telescope looked at the center of the galaxy and saw what we see in figure 2.2; a narrow, long, straight jet of gas blows exactly out of the center. Inside it we can see stronger puffs of gas, so whatever produces the jet points in the same direction for hundreds of thousands of years and also produces a variable amount of gas, but does so in what scientists call a highly collimated beam (that is, a very narrow jet of gas).

Astronomers can see from the jet that it almost certainly comes from a black hole. Why?

There are several pieces of strong evidence. What kind of object can produce such a stable and narrow jet of gas for hundreds of thousands of years? One candidate is an extremely heavy gyroscope. What is the heaviest gyroscope of all? It is a black hole with an accretion disk around it, where material from the galaxy is being swallowed progressively. Astronomers estimate that the black hole in the center of M87 must be around 3 *billion* times the mass of the Sun. So, where does the jet come from? The black hole is surrounded by a massive disk of material swirling around it—the accretion disk—which is slowly spiraling into the black hole. A black hole, although it has a seemingly unlimited appetite, suffers from indigestion. When large quantities of material try to enter the black hole, they generate huge quantities of energy. This tries to blow away the material that is falling in, in exactly the same way that the heat of the star's core keeps it "inflated" and stops gravity from making it collapse. As more and more material tries to fall on the black hole, a limit is reached. The black hole gets an attack of indigestion—so much energy is generated that it counteracts the force of gravity and material is blown away from the two poles of the accretion disk. So, as a black hole is so massive and has such stable spin, the axis of rotation will point in the same direction constantly for hundreds of thousands of years, meaning that this jet of material always points the same way.

The jet is moving away from the black hole at very high velocity, which leads to another odd effect. When an object is moving toward us, we see a Doppler shift to the blue. In the same way, when the object is moving close to the speed of light, the emission of light is also concentrated in front of it in an effect known as relativistic beaming. The greater the velocity, the more tightly the radiation is concentrated in front, like a searchlight beam. When we look down the beam, we may see its brightness increase by a factor of 100 because of this effect. In M87 one of the beams is pointing toward us and the other almost directly away, which means that the former is enormously brightened, while the latter is greatly dimmed and thus invisible, so we see only the jet that points in our direction.

Other pieces of evidence also pointed to the likelihood that there are massive black holes, particularly in active galaxies. In many objects we see extremely rapid variations of brightness. X-ray satellites have shown that the brightness of quasars and Seyfert galaxies[11] may vary by a factor of 10 in a matter of a few tens of minutes. Because an object cannot vary more rapidly than the time taken for light to cross it, this means that a tiny region of space, about the size of the inner solar system, is producing an amount of energy equivalent to as much as 10 or even 100 normal galaxies; such an object can only be a black hole.

Most, if not all, galaxies have a huge black hole in the center. This may be a few thousand times the mass of the Sun, but in some galaxies it may be several billion solar masses. Where does such a huge black hole come from?

There are at least two ways that a black hole millions of times the mass of the Sun can occur. One possibility is that early in the life of the galaxy a massive star forms that creates a black hole when it dies. In the crowded central regions of the galaxy, there are huge amounts of dust and gas floating around freely. Any black hole that forms will start to grow as it sweeps up the material. Initially the growth will be slow, but as the black hole grows, its reach and appetite increases and it becomes more capable of swallowing material. The process accelerates rapidly until the supply of free gas is used up and the amount of material falling on the black hole falls to a trickle.

There is a second way that a black hole can appear. When a galaxy forms, the mass of material increases tremendously toward the center. It is possible that the density is so great in the center of the galaxy that material accumulates naturally until there is so much and it is so dense that it collapses into a black hole, which will then go on growing as it accumulates material. In this way, the black hole starts off much larger and thus grows far faster than if it starts with a star.

Black Holes in the Universe

Whether we like it or not, black holes are a fact of life for modern astronomers. Our violent universe has a whole range of objects—normal stars, white dwarf stars, and neutron stars, which are increasingly more compact objects. Black holes are a natural progression of these; stars of quite low mass die quietly as white dwarfs; larger stars die in supernova explosions as neutron stars; and the largest stars of all have the most spectacular fate, collapsing into a black hole.

We can observe many white dwarf stars directly—the companion star of Sirius (Sirius B, or the Pup) and the companion of Procyon are just two examples. Back at the turn of the twentieth century, astronomers frankly doubted that white dwarf stars could exist. That they were quickly forced to accept them as observations showed that they satisfied a key prediction of Einstein's theory of relativity and that their huge force of gravity red-shifted light that left them.[12] For a long time it seemed that no more extreme object existed. Although it was calculated that no white dwarf star could be more than 1.4 times the mass of the Sun—the so-called Chandrasekhar limit—because it would collapse into a neutron star, this seemed to be merely a theoretical possibility. When pulsars were discovered, astronomers were suddenly forced to accept that this theoretical possibility had suddenly and unexpectedly turned into reality. From neutron stars it was a short step to accepting the reality of black holes.

Nowadays, almost no one doubts that black holes are an astronomical reality and have their place in astronomy, helping us to understand the most violent objects in the whole Universe.

SUGGESTIONS FOR FURTHER READING

On the Internet

"The Discovery of Pulsars," by Jocelyn Bell-Burnell
www.bigear.org/vol1no1/burnell.htm

> *The discoverer's personal story given as an after-dinner speech in the mid-1980s. Jocelyn Bell-Burnell is a wonderful speaker with a wry sense of humor, a complete lack of self-consciousness about her momentous discovery, and a total lack of bitterness that somebody else took the Nobel Prize that should have been hers.*

List of black hole candidates
www.johnstonsarchive.net/relativity/bhctable.html

> *A page in which an updated list of black hole candidates, both in our own Galaxy and in other galaxies, is maintained along with some of their basic parameters.*

CHAPTER 3

Who Is the Strangest in the Cosmic Zoo?

By the end of the twentieth century astronomers were starting to bandy around the most remarkable range of names for celestial objects, names that would have left their predecessors of a century before nonplussed. We have already met white dwarfs, pulsars, neutron stars, x-ray binaries, black holes, and supernovae,[1] none of which existed in the astronomical lexicon in 1900. These were as nothing compared to the bewildering array of new terms that littered astronomy in the 1970s, 1980s, and 1990s, when quasars, blazars, quark stars, and gamma ray bursters impinged ever more dramatically on the public consciousness as scientists became aware of objects that were ever harder to understand and explain.

Back in the 1970s Prince Charles inaugurated the new 3.5-meter Anglo-Australian Telescope in Siding Spring, Australia. In his speech, which was, at times, highly amusing, he made reference to the argot of the astronomers there who had evidently been trying to impress him. The prince lamented that on his visit he had become totally bemused by the multitude of terms thrown at him for different types of astronomical objects and the techniques used to study them, some of which he could barely repeat, let alone understand. Contrary to popular belief, astronomers do not deliberately choose these terms to impress and to blind the public with science—although a little showing off is acceptable from time to time. After all, we *are* human, whatever cartoons and the wackier type of film may try to suggest. An inconvenient name—too long or too complicated—will either be abbreviated or rapidly forgotten. A case in point is the term *quasar*. These were discovered as emitters of strong radio signals that, when examined with an optical telescope, looked like stars—hence, the name quasi-stellar radio sources, or sometimes quasi-stellar objects.[2] This was a bit of a mouthful and was rapidly reduced down to quasars. Perhaps it is because so many astronomical objects have short, simple, but ultimately impenetrable names like quasar or blazar that they fire the public imagination: you may not understand what they are, but that does not stop your imagination going into overdrive.

Here we are going to look at two of the more recent and exotic denizens of the cosmic zoo. There is a risk in such discussions; in astronomy, ideas and theories can change fast, and what we think we know today may be proved wrong tomorrow and look utterly ridiculous the day after. So when we talk about the most exotic citizens of the Universe, it is in the uncomfortable knowledge that it may only take one new discovery for all our ideas to change completely and for what is written here to look quaint at best and, in the worst case, absurd—the Universe can be cruel with those who seek to explain it.

The "Silly" Quasar

Back in the late 1980s a chance encounter in the observatory residence in La Palma led me to spend many nights observing with a Finnish astronomer, Leo Takalo, in the recently built 2.56-meter Nordic Optical Telescope. The telescope was not officially open, but as Leo was one of the astronomers testing it and preparing it for operations, I was able to get access to this superb instrument before it was available to other astronomers. At the time I was directing a Spanish student's master's thesis, and the subject we had chosen to study was an object called OJ287, a quasar in the constellation of Cancer. According to various reports, OJ287 could show on occasion periodic variability of its brightness. Because this periodicity hints that something is spinning or rotating, these reports had assumed a huge importance to astronomers hoping to learn some of the secrets of quasars. OJ287, though, over the many nights that we watched it, showed an unfortunate propensity for unusual, not to say utterly bizarre behavior, even by the standards of quasars. This led to a frequent cry in the telescope control room over the years—"oh dear! OJ287 is being silly again!"

Silliness aside, OJ287 is particularly interesting and unusual. It was also one of the first *blazars* to be discovered.

The story of OJ287 starts in 1968. By then astronomers had discovered dozens of quasars, and new identifications were no longer big news. Even if astronomers did not understand what quasars were, they at least thought that they had a good photofit of a typical quasar. The idea was that all quasars fitted the following five-point profile:

1. They looked like stars, but . . .
2. They were much bluer than any normal star.
3. They were strong sources of radio energy at long wavelengths but were progressively weaker at shorter ones.
4. They had spectra with strong, broad emission lines that showed a large red shift.
5. They were variable in brightness, although their brightness changed only slowly over weeks and months.

Within two years astronomers found examples of quasars that broke every single one of these rules and a number of unusual objects that, most inconsiderately, broke most of them.[3] In their time-honored tradition, just when quasars seemed to be coming under some semblance of control, life was about to become interesting again.

OJ287 was identified in 1968 as a magnitude 15 counterpart of an unusual radio source detected by the Ohio State University's "Big Ear" radio telescope in what was, at the time, the most sensitive survey of the radio sky ever carried out.[4] Initially the observations were sporadic, but these showed that the object was wildly variable. Not only did its visible brightness change, by a large amount, but so too did its infrared and radio brightness, far more than most quasars. For a couple of years around 1971, it got enormously brighter (by about a factor of 15). Because OJ287 was not just much brighter in the visible but also in the infrared and radio, and thus was interesting to many different kinds of astronomers, it attracted a lot of attention and, since then, has been a favorite target for many astronomers.

What made OJ287 interesting was a series of things. First, unlike other quasars, the spectrum was blank. There were no strong emission lines; in fact, there was no hint of anything at all in the spectrum. There was nothing whatsoever to hint to astronomers at what distance it might be. In this sense, it was similar to a group of objects that astronomers had started to call BL Lac objects, or Lacertids,[5] for the first one of its kind to be recognized, but OJ287 was more extreme.[6] OJ287's brightness was so changeable that it could vary by a large amount in a short time. One bewildered group of astronomers was looking at it one night with the 3.8-meter United Kingdom Infrared Telescope (UKIRT) at Mauna Kea (Hawai'i) and saw, to their astonishment, that it brightened by a magnitude (that is, by a factor of 2.5) in just 50 seconds.[7] Astronomers realized rapidly that when they observed OJ287, life was rarely boring.

By the mid-1970s astronomers knew that OJ287 and several similar objects belonged to a new class of beast in the cosmic zoo. These objects showed extreme behavior: rapid and violent variations occurred in all ranges of the electromagnetic spectrum; their spectra were featureless or nearly featureless; and, most surprisingly, their light was strongly polarized. This kind of object became known as a *blazar*, although this term was used for the first time only in 1984.[8] The name comes from BL Lac quasar, but also neatly describes their unusual ability to blaze out suddenly with huge increases in brightness.

Most people are familiar with Polaroid lenses, which are now much used in sunglasses. They reduce glare by allowing through only the light that vibrates in one particular direction. When light is reflected, for example, off the windshield of a car, it gets polarized, that is to say, the light waves line up preferentially one way. If we look at that reflection through Polaroid sunglasses, the material in the lenses will block all the polarized light, thus considerably reducing the glare from the reflection. The effect is rather like trying to pass a large cake through some bars; only when you turn it vertical can it slip through—any other orientation and the bars will block it from passing. This is exactly what a Polaroid material does to light.

Astronomers use a calcite crystal, which acts in the same way as a Polaroid lens; as the calcite in the telescope is rotated, an object that emits polarized light will brighten and fade according to the angle of the crystal as the orientation of the light lines up with the orientation of the crystal. The amount that the bright-

Cosmological Enigmas

ness changes depends on how greatly polarized the light is. If the light is strongly polarized, we will see big changes in brightness as the crystal rotates; if it is only weakly polarized, there will be little change.

The results were astonishing. Most quasars are so weakly polarized that the polarization is virtually undetectable. Blazars are highly polarized. In one particular case, the polarization, amazingly, was greater than 40 percent. But the polarization of these objects was also highly variable. The blazar 3C345, in the constellation of Hercules, went from showing essentially zero polarization to 35 percent polarization in only a few weeks. This kind of behavior was bewildering and, although it was proof positive that something unusual and extremely violent was happening, it was impossible to account for what was being observed.

My interest in OJ287 had been awakened by the possible periodic variations in its light curve. Being OJ287, nothing was as simple as it looked.

The first report came from two astronomers, Visvanathan and Elliot, working with the 1.5-meter telescope at Mount Hopkins in Tucson, Arizona (now known as the Fred Lawrence Whipple Observatory). While observing in 1972, they found that OJ287 showed a periodic variation—that is, its brightness increased and decreased rhythmically—within a span of 39 minutes.[9] This report created great interest because it suggested that something inside the blazar was rotating—possibly material spinning around a black hole before being swallowed—and thus that potentially valuable data could be obtained about the conditions in the center of this blazar from studying the periodicity. The variations were tiny, much less than 1 percent of the brightness, but they did seem to be very much present. Two other groups published similar studies shortly afterward, and, in the grand tradition of OJ287, the waters started to muddy. One report confirmed the periodic variations.[10] The other group, observing at almost exactly the same time, saw no periodicity whatsoever.[11] This established a pattern that was to be repeated many times.

What highlighted the behavior of this extraordinary object were two studies published together in the prestigious journal *Nature* in 1985. In them a Finnish group, led by Esko Valtaoja,[12] and a Mexican group, led by Luis Carrasco, both reported that they had found a periodicity in the variations in the brightness of OJ287, the former of 15 minutes mainly in radio data, the latter of 23 minutes in visible data.[13] The fact that two groups in different countries and using completely different techniques had found similar results and published them in such an important and highly visible scientific journal made the issue of periodicity in quasars both topical and respectable. Many astronomers tried to confirm these results, but although some did see periodic variations, they were of different periods; other groups found nothing out of the ordinary at all.[14] Something odd was going on in OJ287 that allowed it to show rhythmical variations at times, but never for very long, and not particularly consistently.[15] For many years now, the periodicity in OJ287 has returned to its former position—well below the level of scientific consciousness—where, no doubt, it will remain, until some other luckless astronomer suffers from this object's little-known sense of humor.

If astronomers thought that the issue of periodicity in the brightness of OJ287 was dead and buried, however, they had another think coming!

Back in 1987 the distinguished Finnish astronomer Mauri Valtonen was directing the Ph.D. thesis of one of Tuorla's many up-and-coming astronomers. This astronomer came to Mauri's office one day showing him a light curve of OJ287 made up of measures of images found on old photographic plates from the Tuorla collection—like many other quasars, OJ287 had been observed many times by accident over the years and these measures are an invaluable record of its behavior. For OJ287, the first images that were found corresponded to the year 1891. There were nearly 200 measures in total until it was officially discovered in 1968 and became subject to regular and detailed monitoring. The student thought he could see something significant in these measures; Mauri recounted to me some years later that his student had suggested to him that the light curve showed evidence of regular fades in brightness that might be eclipses. Mauri looked at the evidence and replied sagely that he could not see eclipses, but he could see what seemed to be a series of regular brightenings or outbursts in the light curve. Looking into the data in more detail seemed to show that about every 11.6 years, over a period of some 90 years, OJ287 had suffered a major crisis and had become much brighter for a few weeks or months (see figure 3.1).

What could cause a quasar to become as much as three magnitudes brighter for a short interval every 12 years and do so in a seemingly regular fashion?

Mauri Valtonen and his collaborators had a thought-provoking idea. What if, unlike a normal quasar, OJ287 had two black holes in its heart? They suggested that one was 5 billion solar masses and that the "small" black hole of 20 million times the Sun's mass orbited around it in a highly eccentric orbit every 9 years. From Earth, we would see this as a period of 11.6 years thanks to the stretching of time caused by relativity. Every 9 years, when the small black hole passed close to the large one, the encounter would cause a massive perturbation of the large black hole. The force of gravity of the small black hole would cause an enormous surge of material to fall onto the larger one from its surrounding accretion disk and, with it, a huge increase in the brightness of the blazar.

Initially the idea took some time to capture the imagination of the astronomical community,[16] but it made one extremely interesting prediction. As we have seen, there was a big outburst in the light curve in 1972. Another had followed in 1983. That meant that a further outburst should follow in 1994; what is more, the group made the prediction that it would be a double one, with two outbursts about a year apart. Making predictions is the other way to test a theory, but it is a double-edged sword: get it right and people will be impressed (but, even then, you may find that there are alternative explanations suggested as to why your prediction has worked!); get it wrong and you both look and feel silly.

In the early 1990s the astronomical community was not ready to take such predictions seriously—with good reason. Over the years, many people had tried to make predictions about the future behavior of different quasars by using periodicity in the light curve. Despite many attempts and many predictions, however, there had been a 100 percent record of failure. Quasars simply did not have regular or predictable behavior, and to make predictions was to guarantee falling flat on your face; in some cases, the prediction was found to be wrong before it had even been published.[17] It was in this atmosphere that I teamed up with the Finns to try and buck the trend.

The best way to attack the problem, we decided, was to look critically at the light curve and satisfy ourselves there was good evidence that every 11.6 years or so there was a major outburst. This we did, but we also saw that there were two unfortunate flies in the ointment. First, the data that we were working with were of a highly variable *quantity* and *quality.* The observations from before 1968 worked out at about one measurement every six months, but that was strictly an average—at times several years would go by with no data and then there would be a whole lot of measures together. After 1968 the number of measurements jumped suddenly and after the 1983 outburst it jumped even more. Just like archaeologists, we were trying to piece together a puzzle where we knew a lot about the present, a fair amount about the recent past, and little about the most distant past—but what had happened in the most distant past was the most important information for us. Added to this, most of the old observations were not very good, while the most recent ones were by far the best; this gives an idea of the difficulty. The second problem was that the intervals between the outbursts were not completely constant—always a worrying sign when you are looking for periodic behavior—and that the interval between the two most recent and best observed in 1973 and 1983 was the shortest ever. After carefully reanalyzing the data, we concluded that there was genuine evidence for some kind of period of 11.6 ± 0.5 years; the results were published.[18] With only two years before the next expected outburst, it was time to act.

We decided it was necessary to gather an international group of scientists to study OJ287 and make plans. The Finns proposed a small island in the Baltic archipelago just off the southwest coast of Finland where the University of Turku has a small research station.[19] The group was a small but highly varied one. There were several Finns from Turku, including Mauri Valtonen; I attended with José Antonio de Diego, my Spanish student; also present were Jochem Heidt, a German Ph.D. student; Sen Kikuchi, a Japanese researcher; and Phil Charles, a British astronomer. We decided it was extremely important to get together a big project to observe OJ287 in great detail around the time of its next outburst. Over the next few months, additional team members were recruited in different countries. The most important of these new team members was an amateur astronomer from Ottawa, Canada. Over the next three years Paul Boltwood demonstrated a phenomenal level of professionalism and dedication. He had bought a 12.5-centimeter refractor, put it in a dome in his backyard, and had built and mounted his own CCD camera on it.[20] Despite living in a suburb of Ottawa and being surrounded by street lights, he obtained huge amounts of high-quality data with this setup, automating his telescope and the reduction of the observations.[21] Throughout the project, the words "Boltwood Observatory" became synonymous with reliability.

The project started in autumn 1993 as OJ287 slid into the morning sky. We were helped by the award of a large block of time on the different telescopes in the Canary Islands as an "International Time Project."[22] History records that the first observation of the campaign was taken by Paul Boltwood on September 20, 1993; more than 3,000 observations were to be taken just in the V filter (and more than 8,000 in total in all the visible filters) by the time the project ended in spring 1995. As the project advanced, many observatories pooled their data, with the result that over the course of the project observations were carried out in a total of 25 different filters or ranges of frequency from the radio to gamma rays.

Would OJ287 Cooperate and Do Its Stuff?

One of the problems was that nobody knew what to expect. *Predictions* were plentiful, most in happy disagreement, but nobody knew which, if any, would be correct. The best prediction gave a probable date of mid-November 1994 with a second outburst about 13 months later, but nobody was sure how much any prediction could be trusted. The data in figure 3.1 also suggested that the size of the outbursts varied with a longer cycle of some 60 years: the outbursts in 1913 and 1972 were especially large, and after 1913 they became much smaller again before getting progressively larger, while in 1983 the outburst had been much smaller than in 1972. It looked like 1994 would come near the bottom of the cycle, so that the outburst would be quite small and then, after 1994, successive outbursts would get bigger and bigger. Another especially large one is due around 2031.[23] If a small outburst really did happen in November 1994, we thought we would be onto something.

Early in December 1993 I was observing with the 82-centimeter Telescope at Teide Observatory in Tenerife. That autumn OJ287 had been unusually faint— below magnitude 16. On the morning of December 6, 1993, it was much brighter, a full two magnitudes brighter than it had been two months previously. Was this the big outburst? Less than an hour later, a German group observed it and found that it was little more than normal brightness. Had I made some dreadful mistake? When the data were compiled, we found that OJ287 had faded by about a magnitude in half an hour—equivalent to switching off about a trillion suns. OJ287 was being silly again. In fact, it was a small, precursor outburst. It had lasted for about a month, and at the peak of the outburst there were some violent variations in brightness. It had been my good (or bad) fortune to be observing exactly at the peak of one of these violent oscillations of brightness.

As more data came in through early 1994, something else started to appear. OJ287 was brightening slowly but persistently. Could this be the big outburst?

As we watched, the brightness of this blazar increased all through 1994, until we lost it behind the Sun on June 13. It was still fainter than it had been briefly in

December, but it was going in the correct direction. What would happen during the nearly three months that it would be too close to the Sun in the sky to be observable? Might the outburst happen without us ever seeing it?

On September 8 the amazing Paul Boltwood recovered OJ287 low in the dawn sky in Canada—far too low for any professional telescope to observe. There was some consternation when we realized that it was much fainter than it had been in June. Over the next three weeks it brightened constantly and, by the end of September, it was obvious that something interesting was happening. Through October, OJ287 continued to brighten until, lo and behold, on November 11, 2004, it broke fractionally through magnitude 14 for a few hours before fading rapidly again. What was more, the maximum was the faintest that had ever been seen, with the possible exception of 1924 when the data were very poor.

It seemed that the predictions of the binary black hole model had been amply confirmed. If the second outburst happened late in 2005, we would be overjoyed.

By April 1995, OJ287 was as faint as we had ever seen it. Just before disappearing behind the Sun again, though, it was brightening. Thirteen months after its first outburst, on December 16, 1995, OJ287 reached an almost identical magnitude (see figure 3.2). Our cup runneth over.

Astronomers finally started to take the binary black hole model seriously. For the first time, genuine predictions had been made about a quasar's behavior, and they had come true. The next outburst was due in mid-2006, but there was a slight problem. One might think that with several outbursts of OJ287 well observed, it would be possible to predict with great precision the orbit of the binary black hole and the time of the next outburst. Unfortunately, it is not. By making tiny changes in the assumptions, a wide range of predictions come out when the data are fed into a computer, and it is asked to model the two black holes. The prediction for 2006 was uncertain by as much as six months.

In fact, since the original version of this chapter was written, according to amateur data from a large international group of astronomers led by the former director of the British Astronomical Association's Variable Star Section, Gary Poyner, the outburst actually took place early in November 2005 and was, as expected, rather larger than the famous 1994 outburst. This result has been somewhat disconcerting and is still being analyzed. Recently I gave a couple of talks with Gary presenting the data from the observing campaign. Mauri calculates a revised mass of the large black hole of 19 billion solar masses, rather than the previous value of 17 billion—Gary looked rather bemused when I asked him if he had ever added 2 billion solar masses to a black hole.[24]

OJ287 has been trumpeted as an example of true periodicity in a quasar. But is it?

If OJ287 is really periodic, after a century of observations we should be able to measure the interval between its outbursts better. Could the problem be more sinister?

Suppose the light curve of OJ287 shows a period of exactly 11.6 years. That means that if we slice up the light curve of OJ287 into pieces 11.6 years long and stack them on top of each other, the outburst should happen in exactly the same place every time. If the period is a little longer than that, they will appear to happen slightly earlier each time; if shorter, they will happen slightly later. As figure 3.3 shows, they line up just nicely if you assume that the period is 11.85 years. So, what is the problem?

The problem is that you can only get them to line up in 1948, 1960, 1983, and 1994 and, just possibly, 1972, but there are too few data to be certain. Before 1948 there is no evidence of the period whatsoever, mainly because nobody was looking at the right times, although in 1912 people *were* looking just about the time when OJ287 should have been especially bright, and it was not (that *could* be down to bad luck and the maximum being brief and just missed). There are two maxima separated by a little over a year. The second maximum lines up neatly in 1995, 1984, 1937, and (probably in) 1925 and 1913, but it is *not* there in 1901 and 1973. In other words, there is just enough doubt to make me wonder if OJ287 was just fooling us.[25] We will see in late 2007. As I revise this text in April 2007, we are watching, enthralled, as OJ287 starts to rise steadily in brightness once more. Mauri has been very definite in his prediction this time, and as the months advanced and OJ287 continued to fade from its last outburst, without showing any signs of starting to brighten again, we all started to get a little nervous. Right now, exactly five months and two days before what Mauri calculates should be the peak of the outburst, it finally seems that OJ287 is waking up and smelling the coffee. However, this new outburst seems so discrepant with a constant 11.85-year period, coming as it has more than a year too early, that there seems to be no question that it has killed all talk of a regular periodicity DEAD!

What Makes a Blazar Like OJ287 Special?

Even if astronomers have a hard job explaining the singular behavior of blazars, they have a good idea what makes them so peculiar. OJ287 is unusual for two reasons, one of which is the presumption that it has two black holes in its heart rather than the single one in most quasars and blazars. The second, though, is a product of the black hole.

Black holes do not have an insatiable appetite. If more material tries to fall into one than it can swallow, the excess will be expelled at almost the speed of light in two jets that will blow out from the poles of the black hole's accretion disk. Normally we do not see this jet, but what happens if it points toward us? Just by luck, a few quasars will happen to line up such that the jet points almost exactly toward the Earth. This is what leads us to see a quasar as a blazar; we are looking more or less along the jet. Inside it, material is moving toward us at more than 99.9 percent of the speed of light, giving rise to some odd effects from relativity: first, the light is concentrated like a lighthouse beam—the faster the material in the jet is moving, the more concentrated the beam. This means that we see the blazar as much more luminous than it really is. Yes, a searchlight *is* bright, but what makes it seem so intense is that reflector behind it that concentrates the light into a tight beam; that is what makes a blazar seem brilliant. It also means that everything that happens in the blazar appears accelerated: it seems to move faster and happen much more quickly than it does really; blazars appear to vary quickly and violently because the jet magnifies *everything* that happens inside them.

OJ287 is special because its jet is particularly well lined up with the Earth—it is off by only about 13 degrees. This makes it one of the best-aligned jets of any known blazar. OJ287 is also relatively nearby, which means that by being so well aligned with the Earth and being so close, we actually look some way inside the throat of the jet. That is why this object is a particularly extreme blazar that behaves in such an unusual and bizarre manner. Other blazars are gentler in their behavior than OJ287 because their jets are less well aligned with the Earth and so we are only in the fringes of the intense beam of light.

I am confident that the second outburst predicted by Mauri Valtonen's binary black hole model will genuinely happen in 2007. However, the theory that there is a genuine periodicity in OJ287's behavior died with that early outburst in 2005 because it cannot be fitted to any period, however tricky the math that is used. The old data may be equivocal, but the new data brokes no argument. What is certain, though, is that OJ287 is one of the most curious and bizarre objects in the known Universe.

The Wrong-Way Bang

One of the most curious cases of accidental discovery revealed probably the strangest and most mysterious objects in the Universe. Certainly they have taken the prize from the blazars as the most luminous and violent objects in the cosmic zoo. What was oddest is that the discovery was made by the military, who had hoped not to see anything at all! If that sounds confusing, pity the poor military scientists who were in the midst of the affair.

In 1963 the United States and the then Soviet Union signed the Test Ban Treaty, intended to stop atmospheric tests of nuclear weapons. To support this, the United States launched a series of satellites called Vela—from the Spanish word meaning "to watch over."[26] The aim was to detect the flash and the radiation from atmospheric nuclear tests. In 1969 the initial set of what was intended to be ten Vela satellites (they were so successful that only the first three pairs had to be launched, with the final two pairs not needed) was replaced by two pairs of more sophisticated satellites—the advanced Vela satellites, or Vela 5A & B and Vela 6A & B. These satellites were part of a program run jointly by the Advanced Research Projects of the U.S. Department of Defense and the U.S. Atomic Energy Commission, managed by the U.S. Air Force. They had both x-ray and gamma ray detectors on board and electromagnetic pulse sensors and went into especially high orbits (100,000 kilometers)[27] that allowed each of them to see an entire hemisphere of the Earth and to be well above the effects of the Earth's atmosphere.

In all, the Vela satellites detected 41 nuclear tests but, within weeks of starting to operate, the new Vela satellites also generated alerts of possible nuclear explosions that severely disconcerted their Air Force controllers. The results were so unexpected that they were kept strictly secret for some years. Between July 1969 and April 1979 the four advanced Vela satellites detected 79 brief but powerful bursts of gamma rays that did not seem to come from the surface of the Earth. This initially raised the frightening possibility that a country or countries might be carrying out nuclear tests in deep space to get around the Test Ban Treaty. If so, there were some extremely unpalatable military possibilities of untraceable and unreachable weapons in space, upsetting the delicate balance that had made war unthinkable.

By 1973 scientists at the Los Alamos National Laboratory had proved that the bursts of gamma rays came from deep space and that they had a natural origin. However, no progress at all could be made in identifying the sources of these mysterious bursts of radiation. The problem was twofold. First, the bursts of gamma rays were brief. Second, the detectors used in the satellites had such poor resolution that it was impossible to pinpoint the sources of the gamma rays sufficiently well to direct an optical telescope at them and identify their source. When astronomers looked in the general area of sky where the outbursts came from, there was no obvious culprit. There the problem remained for many years. Astronomers knew that what they had termed "gamma ray bursts" must come from outside our solar system, but knew no more than that.

Finally, in April 1991 the Space Shuttle Atlantis carried into orbit a new gamma ray satellite called the Compton Gamma Ray Observatory. A fundamental part of this satellite was an experiment called BATSE (Burst and Transient Source Experiment), which was designed to detect brief bursts of gamma rays from space. BATSE used eight detectors to cover the whole sky with a sensitivity vastly greater than any previous satellite. More important, it could detect the position of the outbursts with a precision far greater than previously possible, to within the diameter of the full Moon. BATSE was able to produce a map of where in the sky the gamma ray bursts, or GRBs came from. In its nine years of operation some 8,000 GRBs were detected, most of which lasted no more than 20 seconds.

BATSE was a massive improvement on anything that had gone before, but was ultimately frustrating because the positions were still not quite good enough to allow astronomers to identify a GRB. BATSE became torture. At the time, I was working as a support astronomer at Teide Observatory in Tenerife, assisting visiting astronomers with their observations. On a regular basis, agitated telephone calls would arrive during the night telling us that BATSE had found another GRB and that the position was about such and such and could we *please* take a look and see if there was anything odd in that position. Doing a favor for a colleague is an integral part of the way observatories run, and most astronomers are willing to help out. After two or three such alerts, though, it became clear that however hard we tried to help and no matter how precisely the astronomer who called had pinned down the position of the GRB, it was never going to be possible to identify the source. It was frustrating!

BATSE did give astronomers one invaluable piece of information: the GRBs were spread evenly around the whole sky. This means they are either extremely close in space or extremely distant. The reasoning is simple. Suppose that they were produced inside our Galaxy; as our Galaxy is a rather flat disk, we would expect to see the GRBs distributed around the Milky Way with only a few in other parts of the sky. If the GRBs are produced in the center of the Galaxy, we would see them all concentrated around this point in the sky in the constellation of Sagittarius. Only if the GRBs are produced within a few tens of light years of the Sun, within the thickness of the disk of the Milky Way, would we expect to see them evenly distributed all around the sky.

It was hard for astronomers to imagine how huge bursts of gamma rays could come from thousands of different points in the space around the Sun. They would have to come from normal stars, and normal stars do not emit sudden bursts of hard gamma rays. The alternative was that they came from extremely distant galaxies. Why? The reason is the same. Suppose the GRBs come from galaxies only a few hundred million light years away. We know, though, that such relatively nearby galaxies are not randomly distributed in space; there are far more in the direction of the constellations of Virgo and Coma where what is termed our local supercluster of galaxies is centered. But we do not see more GRBs in the direction of Virgo and Coma. Ergo, they must come from far more distant regions of space.

There was a dramatic breakthrough in 1997. One year earlier, a Dutch-Italian satellite called Beppo-Sax was launched. This satellite had both x-ray and gamma ray telescopes on board. Many gamma ray bursts were known to emit some x-rays too, and x-ray telescopes can form much sharper images than their gamma ray counterparts.

The first significant event was on February 28, 1997. Beppo-Sax detected a GRB that was cataloged as GRB 970228 (for the date). The x-ray detectors found a new source close to the best estimate of the position of the gamma ray burst and tied down its position accurately. For a few days it seemed that nobody was looking at any other place in the sky. On March 1 radio astronomers at the Very Large Array at Socorro, New Mexico, detected a faint radio source that seemed to be from the gamma ray burst and were able to give a precise position that allowed astronomers to check if there was anything unusual there, even if it was extremely faint. Within a few hours of the radio observation being announced, a faint fuzzy object had been found at that position in images taken with the 3.5-meter telescope at Apache Point Observatory in New Mexico. Astronomers at the Multiple Mirror Telescope on Mount Hopkins examined the spectrum of this object and found that it was, as suspected, a faint galaxy with a red shift of around 0.5.

If this was the source of the gamma ray burst, it was at a huge distance. But was it? A group of Italian astronomers thought they could detect a tiny movement of the faint afterglow from the explosion in two Hubble Space Telescope images taken less than two weeks apart in late March and early April. If the source was really displacing itself detectably in the sky in such a short time, it had to be extremely close in space, certainly within our own Galaxy. Just when astronomers thought that they had started to make huge progress in understanding this annoying object, a sizable wrench had been thrown in the works. Finally, new images taken with the Hubble Space Telescope in September showed that the object had not really moved at all and that six months later it was in exactly the same place as before; the collective sigh of relief from astronomers was enormous.

By now, other events had taken over, and the pace of discovery was picking up. On May 8, 1997, Beppo-Sax detected a GRB that was cataloged as GRB 970508. The x-ray detectors tied down its position accurately, and an alert was sent out over the Internet. Within seven hours one of the small telescopes at Kitt Peak National Observatory in the mountains above Tucson had tracked down the faint afterglow from the explosion in the constellation of Camelopardus as it faded. This time there was no doubt; it was like police following a felon escaping a crime scene in his own car—they might have missed him committing the actual crime, but he was not going to get away. Three nights later the glow was observed using the 10-meter Keck Telescope on Mauna Kea in Hawai'i, which analyzed its spectrum to detect a red shift and thus measure its distance. The result was not unexpected but still astonishing—the galaxy where the GRB had appeared had a red shift of 0.835 and thus, like the suspect galaxy for GRB970228, it was at an enormous distance, certainly more than 6,000 million light years, and hence the quantity of energy released in the outbursts had to be phenomenal.

What was missing, though, was to catch one of these explosions in the act. As most of the GRBs lasted no more than 20 seconds, this was going to be extremely difficult as the best reaction time astronomers had managed was to look a few hours after the outburst had happened. Astronomers, though, had been helped by what, at the time, had seemed to be a major disaster.

In 1992, the Compton Gamma Ray Observatory, still in the early days of its mission, suffered a major equipment failure. Initially all the data were recorded onto tape and stored on two on-board tape recorders for later transmission to Earth. This meant that each time that BATSE detected a gamma ray burst, the information remained in limbo until the controllers on Earth had contacted the satellite and ordered it to play back its data, something that was done just once a day. With no on-board storage capability, they had to plug the satellite into the real time data network run by NASA to transmit data to Earth as it was taken. At the same time the Internet was also starting to become widespread, and the mission scientists decided to take advantage of this. Data from the satellite were intercepted at Goddard Space Flight Center in Maryland, and data from BATSE were removed from the stream. These data were then processed rapidly, and an approximate position for any GRB calculated; within four seconds the news was spread around the Internet to any interested observer. It was these Internet alerts that were the source of the messages that we were receiving so often at Teide Observatory. All that was missing was a way to use this information profitably.

The solution took seven years to arrive, but it was worth it. In 1999 an instrument called ROTSE (Robotic Optical Transient Search Experiment) started to operate at Los Alamos in New Mexico (see figure 3.4), and was to detect the visible flash from GRBs as they happened. This was an incredibly simple idea. As the positions of GRBs generated by BATSE were so poor, ROTSE consisted of four ordinary 200-millimeter telefocal lenses on a computer-driven mounting. The cameras gave a field of view of about 16 degrees—that is more than 30 times the diameter of the Moon, equivalent to the error in the position estimated by BATSE. When an alert of a GRB was received the cameras could be swung to the target area of the sky within seconds and start taking images.

On January 23, 1999, ROTSE's computer received an alert and swung the telescope into position. Twenty-two seconds after the gamma ray burst started, ROTSE was taking images and, when they were analyzed, a magnitude 9 star appeared suddenly and then faded. What was interesting was that the flash of light was slightly delayed compared to the gamma ray burst, so that when ROTSE started to observe, it was still brightening. It was also interesting that the alert itself had been generated by a small, preliminary flash of gamma rays that lasted about 7 seconds, at which point a gigantic burst—one of the 2 percent of brightest outbursts ever observed—started, taking less than 2 seconds to reach maximum and start to fade; just 20 seconds after the original faint flash, the GRB was over while the optical flash was just starting.

With the position pinpointed so precisely, many telescopes could observe the afterglow of the outburst, among them the famous Schmidt camera of Mount Palomar, in California, which observed it four hours later, when the flash had already faded to magnitude 18. Later on, the 2.56-meter Nordic Optical Telescope in La Palma (Canary Islands, Spain) and the 10-meter Keck Telescope on Mauna Kea also observed it, as did the Hubble Space Telescope (see figure 3.5). The spectrum of the fading afterglow showed a red shift of 1.6, meaning that the GRB had occurred at a distance of 9,000 million light years. The brightness of the optical flash was an amazing 10,000,000,000,000,000 times greater than the Sun—10,000 times the luminosity of the entire Milky Way. Had it occurred at a distance of 3,000 light years, the optical flash would have been as bright as the Sun. This led some people to link gamma ray bursts to the extinction of the dinosaurs: a nearby GRB might have killed the dinosaurs by frying them with radiation. To back up the theory, it has been suggested quite seriously, and with good reason, that the less famous Ordovician extinction 450 million years ago (when most life was confined to the oceans and thus protected from many potential catastrophes) could well have been caused by partial destruction of the ozone layer produced by a nearby massive outburst.[28]

Other GRBs were observed that offered interesting clues. GRB 990510 was observed with many telescopes in the Southern Hemisphere, particularly the big new instruments in Chile. The spectrum of the afterglow, even though it had already faded down to magnitude 25 by the time that the telescopes could observe it, showed many absorption lines of metals superimposed on the rainbow spectrum of hot, excited gas that allowed astronomers to measure its red shift as being 1.6. This GRB, though, also offered astronomers the chance to carry out some special measurements that had never been tried previously, observing polarized light from the afterglow of the explosion.

The GRB afterglow was observed to be weakly polarized. This was a huge clue about the way its energy was produced. Just 1.7 percent of the light was polarized, but this was far more than would be expected from a hot gas. Apart from when light is reflected, polarized light is also produced in objects such as quasars and supernovae where there are strong magnetic fields and lots of electrons flying free in space at enormous velocity.[29] This observation suggested that the GRB must be somewhat similar to a supernova explosion, with a very violent event happening in the object (although that last was obvious anyway).

So, what is capable of producing such a gigantic flash of light?

Astronomers are looking at two possible explanations. One is that GRBs may be caused by the collision of two neutron stars making a black hole. The other theory suggests that GRBs may be a hypernova.

Astronomers are certain there are black holes because there are invisible objects in the Galaxy that are more than three times the mass of the Sun. A neutron star cannot grow to be more than three times the Sun's mass because its force of gravity will be so great that the neutrons in it will be crushed into a black hole. So, if two neutron stars collide—each of which are less than three times the Sun's mass, but sum more than this when they come together—the result will be a big flash of energy and a violent collapse into a black hole. GRB observations give a big clue here. Many are in young galaxies where huge numbers of young, massive stars are being formed, but there will be few neutron stars. For two to collide, they would have to form independently in a binary system and slowly spiral into each other. This is not impossible, but the other theory, that of the hypernova, is more probable and attractive.

So, What Is a Hypernova?

In our Galaxy there are at least three exceptionally massive stars. For many years the heavyweight champion of the Milky Way was Plaskett's star. This is a binary system with two stars, each 55 times the mass of the Sun, orbiting each other. It was supplanted by the star Eta Carinae in the Southern Hemisphere. Many astronomers suspect that this star, which is more than 100 times the Sun's mass, is a supernova about to happen. The current record holder is a star in Sagittarius that was discovered only in 1990. It has been nicknamed "The Pistol Star" because it is associated with a nebula that has the distinctive form of a handgun and thus has been named, with breathtaking originality, the Pistol Nebula (see figure 3.6). The Pistol Star may be as massive as 130 times the mass of the Sun. A star like it will live fast and die young. What will happen when such a massive star dies is uncertain, but astronomers think it might become a hypernova and that this, aided by relativistic beaming, might be observed as a GRB from somewhere in the Universe.

According to this theory, a hypernova is an extremely massive star, larger than 25 times the mass of the Sun, or a type known as a Wolf-Rayet star; these stars are extremely rare in our Galaxy. Before it dies, it loses most of its outer atmosphere of hydrogen. The remaining star is still so massive that when it finally runs out of nuclear fuel and collapses, it cannot form a neutron star. In fact, *as* it collapses, the core turns into a black hole. We then have the strange scenario of a star that, on the outside, appears totally normal, but that has a growing black hole in its center. Material continues collapsing onto the black hole in its core at a great rate, and the excess of material that the black hole cannot swallow is blasted out of the two poles of the accretion disk of the black hole at almost the velocity of light, straight through the collapsing outer layers of the star. This jet of material creates a massive pulse of gamma rays that are beamed in the same direction as the jet, concentrating all its energy in a highly concentrated shaft of light like a lighthouse beam in just the same way that we see blazars much brighter than they are really. The result is that a GRB can seem brilliant from thousands of millions of light years away if we are caught in its beam, whereas from much closer it can be completely invisible if seen from one side. Given that we know that the core of a massive star will collapse in just a few seconds, we can understand why gamma ray bursts are so brief and so violent.

The proof—or, at least, strong circumstantial evidence—came from a burst seen on March 29, 2003. GRB030329 was detected the night after its outburst and found to be the closest gamma ray burst yet observed, with a red shift of 0.1685 and thus a distance of "just" 2,650 million light years.[30] The object was observed two nights later with the Kueyen 8.2-meter telescope of the European Southern Observatory in Cerro Paranal (Chile). Using the ESO's Very Large Telescope, astronomers were surprised to see the distinctive spectrum of a luminous supernova. Over the following weeks the object's spectrum showed the characteristic behavior of a supernova. The chances of a supernova appearing in the same galaxy as a GRB within two days of the outburst without the two being connected are so remote as to be almost zero. It is practically certain they are different elements of a single explosion, with the hypernova explosion following the gamma ray burst.

There has been a lot of discussion that gamma ray bursts in our own Galaxy, such as the one that may occur when Eta Carinae or the Pistol Star explodes, could be lethal for life on Earth. However, it is like a searchlight beam: if you are caught in the beam, it is brilliant and you will be brightly illuminated, but you do not have to be far outside it for you not even to notice that it is there. The same thing happens with gamma rays bursts; even though you see the supernova (or hypernova) explosion, you have to be in precisely the right—or wrong!—place to see the burst of gamma rays. It is extremely unlikely that even if the Pistol Star, or Eta Carinae, were to explode as a hypernova, we would be affected by the beam of gamma rays from it.

SUGGESTIONS FOR FURTHER READING

More Advanced Reading

A. Sillanpaa, S. Haarala, M. J. Valtonen, B. Sundelius, G. G. Byrd, "OJ 287-Binary Pair of Supermassive Black Holes," *Astrophysical Journal* 325 (1988): 628–34
Available on the Internet as a PDF file at http://cdsads.u-strasbg.fr/cgi-bin/nph-iarticle _query?1988ApJ...325..628S&data_type=PDF_HIGH&type=PRINTER& filetype=.pdf

Harry J. Lehto and Mauri J. Valtonen, "OJ 287 Outburst Structure and a Binary Black Hole Model," *Astrophysical Journal* 460 (1996): 207–13
Available on the Internet as a PDF file at http://cdsads.u-strasbg.fr/cgi-bin/nph iarticle_query?1996ApJ...460..207L&data_type=PDF_HIGH&type=PRINTE R&filetype=.pdf

> *Neither of these two articles is easy reading, but the former in particular has no complicated mathematics and a good description of the binary black hole model that is not difficult to follow. They provide a fascinating insight into how the idea that the blazar OJ287 has two black holes has developed. These are just two of the articles that have been written by Mauri Valtonen and his colleagues describing the binary black hole hypothesis and that show how the model has changed and been adapted over the years in a response to new data.*

On the Internet

The GRB 990123 Page
www-int.stsci.edu/~fruchter/GRB/990123/

> *This page offers results on this famous gamma ray burst including an animation showing the burst of light fading with time over the year after the outburst. It is not easy reading, but the images are plenti-*

ful and detailed. It is sobering that in the first image of the animated sequence (taken 16 days after the GRB happened), it was already magnitude 25.4 and 4 million times fainter than when the outburst first happened. In the second image it had faded to magnitude 27.7.

ROTSE
www.rotse.net/summary/

This site, a layman's guide to ROTSE's work, includes a superb animation that shows exactly how a GRB changes the appearance of the gamma ray sky. It gives a clear and lucid explanation of studies of GRBs and something of the history of their observation.

www.eso.org/outreach/press-rel/pr-2003/pr-16—03.html

This ESO press release explains the observations of GRB030329, the afterglow of which was later designated as the supernova SN 2003dh. This press release is not particularly straightforward reading, but contains a number of images and spectra and explains in detail the importance and significance of the observations.

CHAPTER 4

How Far Is It to the Stars and Will We Ever Be Able to Travel to Them?

P eople often say "it's all relative," harking back to Einstein. It is a clever phrase to impress your friends, but it is also relevant to life and human experience. Nowhere is this truer than for distances and travel between the stars. How often do we grouse about having to walk a couple of blocks to the shops and how long it takes to get there? Yet we are willing to sit in the cinema and believe that Luke Skywalker or James Kirk can hurtle from star to star between one scene and the next.

Even scientists cannot comprehend the vast distances between the stars. If we use any unit of distance that we are familiar with in daily life, like kilometers, the distance to the nearest star has so many zeros on the end that we cannot take in what it means. Astronomers get around this by measuring distances in "light years" and "parsecs." Even then, when we start talking about the distance to all but the nearest galaxies, we need to use "Megaparsecs" (millions of parsecs) to stop the zeros building up too far.

On this stage of our cosmic tour, the first step is to understand how big our Galaxy is and how great are the distances between the stars. We are then in a better position to talk about the possibility of traveling between the stars. As we will find out, even though science fiction, television, and cinema are littered with star-farers, the practicalities of interstellar travel are far from obvious. Despite the best that science can imagine—time dilation, antimatter engines, and possibly one day even a warp drive—there are many practical problems with interstellar travel, some well known, others less so.

So, let us first take a look at how we have measured first the size of our solar system, then the distances to the nearest stars, and finally the size of our Galaxy. Once we have established the distances between the stars, we can look at the problem of traveling between them.

From Earth to the Moon . . . and Beyond

Measuring the size and scale of the Universe has been an enduring problem for scientists and philosophers. At times it has had a practical element—navigators wanted to know what was over the horizon—and to know how far away a place was and how long it would take to get there; at other times, though, it has just been a product of philosophical musing when we have looked across the ocean toward the horizon and wondered how large Earth is, or looked up at the ocean of stars in the heavens and wondered how far away they are.

The ancient Greeks wondered about such questions. Although hampered by a system of numbers that made Roman numerals look positively user-friendly and that meant that arithmetic and mathematics were so difficult as to be almost unknown in ancient Greece,[1] Greek astronomers made astonishing advances in understanding the scale of the Universe. Several hundred years before the Christian era, Greek philosophers and astronomers estimated the Earth's size and the distance to other bodies such as the Sun and the Moon, thus providing a first idea of the size of our Universe.

In 276 B.C. one of the greatest of the Greek astronomers, Eratosthenes, was born in Cyrene in what is now northern Libya. In about 240 B.C. he became the third librarian of the renowned library of Alexandria. Although Eratosthenes made huge contributions to different fields of mathematics, geometry, geodesy, and astronomy, his contemporaries were less than flattering about him and regarded him as a jack-of-all-trades but master of none. It is a sad reflection of the way that sci-

Cosmological Enigmas

ence often mixes personalities and ability that it has taken more than 2,000 years for him to receive his due recognition.

What is now regarded as his crowning achievement was his famous measurement of the circumference of the Earth, which, for the first time, gave mankind a tiny inkling of how huge the Universe is. Although his work "On the Measurement of the Earth" in which he explained his results is lost, various other authors gave at least some of the details of his calculation.

Eratosthenes compared the length of the noon shadow at midsummer in Syene (now Aswan in Egypt) and in Alexandria, knowing that they were a considerable distance apart and that the shadow cast by the Sun at midday was a different length in the two towns. Knowing how far it was between Syene and Alexandria and assuming correctly that the Earth was round, he could use the different angles that the Sun made (the different lengths of shadow on the ground) to calculate how many times bigger the circumference of the Earth was than the distance between Syene and Alexandria.[2] He gave the result as 250,000 stadia.

Our problem is that no one really knows how long a stadium was. Some scholars have argued that the stadium was 157.2 meters. If this is true, Eratosthenes calculated a value of 39,300 kilometers for the Earth's circumference. Other scholars suggest that the stadium was 166.7 meters long, and so his value was 41,700 kilometers and thus not quite so good (the correct value is 40,075 kilometers).[3] At a time when any land journey had to be made on foot or by horse, Eratosthenes showed how large the world is and how tiny we are in comparison. A less-well-known part of the same calculation gave the first hint that even the solar system is vast compared to the size of the Earth; he assumed that the Sun was so far away that its rays were essentially parallel and thus hinted that it was probably millions of kilometers away from the Earth.

Later, Eratosthenes showed just how tiny the Earth was. By observing eclipses, he estimated the distance to the Sun and to the Moon, obtaining 804,000,000 stadia (124 million kilometers) and 780,000 stadia (123,000 kilometers) respectively. These values were less accurate than his measure of the size of the Earth, but his distance from the Earth to the Sun (what astronomers call the Astronomical Unit) was only 17 percent too small, which is astonishingly accurate—far better than anyone else achieved for nearly 2,000 years. At a stroke, the 250 kilometers from Athens to Sparta became a millionth of the size of the Universe. Even so, the Greeks did not try to measure the distances to the stars and were not to know that they were a million times greater still.

From Crystal Spheres to Distant Suns

Between the second century B.C. and the sixteenth century, Western astronomy did not just stagnate; to a large degree, it went backwards. Much of what the most advanced of the Greeks had discovered was lost. Ideas of a flat Earth held sway. When Columbus tried to reach India by sailing west instead of east, his crew feared for their lives and were close to mutiny, believing that their ship would fall off the edge of the Earth. A system advanced by Ptolemy in which the Sun and planets were held on crystal spheres inside a sphere on which the stars were painted, with all of them rotating around the Earth, had become the dogma. It was so well established that to doubt it risked the Inquisition and death at the stake, a fate suffered by Giordano Bruno as late as 1600 and only narrowly avoided by Galileo some ten years later.[4]

The death knell for the Ptolemaic system, with the Earth at the center of the Universe, sounded early in the seventeenth century. Tycho changed the face of the Universe, although it would take 200 years to remove the last vestiges of the Earth-centered ideas of Ptolemy. Tycho's student, Johannes Kepler, took his mass of careful measurements of the movements of the planets and discovered that, to his considerable surprise, whatever he did to them, he could not get them to fit a circular orbit.[5] Nor could he get them to fit an orbit of "circles upon circles." Because the perfection of the circle had been a fundamental building block of the structure of the Universe for more than two millennia, Kepler's discovery was a profound shock. Between 1609 and 1621, Kepler published his three laws of planetary motion and proved once and for all that the planets orbited the Sun and not the Earth.

With his laws, Kepler made it possible to calculate the distances to the planets with great exactitude. He showed that the square of the time a planet took to orbit the Sun was proportional to the cube of its distance from the Sun. Mars, which was one and a half times as far from the Sun as the Earth, took $1.5^{3/2} = 1.8$ years to complete one orbit, whereas the most distant planet, Saturn, took 29.5 years to complete an orbit and was thus 9.5 times as far from the Sun as the Earth is ($9.5^{3/2} = 29.5$). By knowing the distance from the Earth to the Sun, you could

Cosmological Enigmas

work out the distance to any planet in the solar system. In one fell swoop Kepler had made the Universe ten times bigger than even Eratosthenes had imagined.

How far was it, though, to the stars? Once again, the distance from the Earth to the Sun played a critical role. By the eighteenth century astronomers were beginning to wonder seriously about this problem. A simple thought experiment showed that the stars were probably many times further than even Saturn—the most distant planet then known. Take the Sun. Now, imagine moving it away from the Earth until its brightness is the same as that of a normal star. How far away would it have to be? The answer, although it was not really even guessed then, was around half a million times further away; astronomers at that time knew that if the stars were suns like our own, they would have to be very distant.

The first astronomer to make a serious attempt to measure the distances to the stars was the Reverend James Bradley, in 1728.[6] Bradley used a special telescope mounted in Kew in West London to measure the position of the brightish star Gamma Draconis, which passes almost exactly overhead from the latitude of London. His aim was to measure the parallax of the star—in other words, how much it shifted in the sky when the Earth moved from one side of the Sun to the other. Parallax is extremely simple to understand. Close one eye and line up a finger held at arm's length with a book or some other object on the other side of the room. Now change eyes, and you will see that your finger appears to have jumped to one side. You are now looking from a slightly different angle and see things with a different perspective. Knowing the separation between your eyes and how much your finger appears to move, you can measure the distance to your finger accurately.

Bradley hoped to use the 300 million kilometers between the Earth's position on one side of the Sun and the other to give him two different perspectives viewing Gamma Draconis and thus to measure its distance. To Bradley's dismay, not just Gamma Draconis but also every star that he observed moved the same amount in the sky. He was seeing the *aberration of light*—the fact that light has a finite velocity and the Earth is moving at a certain velocity in its orbit. The effect is the same as in sailing: if the wind is blowing in one direction and you wish to go in a different one, you set the sail at such an angle that the two directions combined push you in the one that you wish to go (it is even possible to make a little forward progress into a headwind this way). Bradley had, by accident, demonstrated that Kepler was right and that the Earth really is moving around the Sun in its orbit. It was also absolute proof of the fact that the speed of light is not infinite, confirming the reasoning of the Danish astronomer Ole Römer that had been published in 1675.

Bradley also demonstrated something else. He could not detect a parallax, which meant that the stars that he was studying had a parallax of less than one second of arc. This is a tiny shift, the equivalent of the diameter of a one-centimeter coin seen from two kilometers away. Bradley was close to detecting parallax, but it would take more than a century for astronomers to learn that.

As we saw in chapter 1, three astronomers, Bessel in Germany, Struve in Russia, and Henderson in South Africa, independently measured the distance of a star at almost the same time in the 1830s. Bessel, the first of them to announce his results, showed in 1838 that the star 61 Cygni had a parallax of 0".29. This meant that its true distance was 710,000 times the distance from the Earth to the Sun (710,000 Astronomical Units), or a small matter of 107,000,000,000,000 kilometers. Both measures have so many zeros on the end that most people simply cannot really comprehend them.[7] Astronomers were forced to use another type of measurement for such huge distances.

Instead, astronomers speak of distances in terms of how long light takes to travel them. Light travels 300,000 kilometers in one second or 9,460,400,000,000 kilometers in a year. Astronomers thus started to speak of distances in terms of light years. Thus the 710,000 Astronomical Units to 61 Cygni became more conveniently 11.3 light years.

Astronomers have not been entirely faithful to the light year, which has a big rival: the parsec. At a distance of 3.26 light years a star would show a parallax of exactly 1 arcsecond. A star with a parallax of one-tenth of an arcsecond is at 10 parsecs (32.6 light years). The star 61 Cygni, with a parallax of 0.29 arcseconds, is at $1/0.29 = 3.4$ parsecs distance. This numerical convenience makes the parsec attractive to astronomers.

Bessel was successful in picking, as he had hoped, a nearby star; 61 Cygni is now known to be the eleventh-closest star system to the Sun. Henderson had done even better, picking the very closest star system to the Sun, which shows a parallax of 0.76 arcseconds and is thus at 4.3 light years, or 1.3 parsecs.

Thus, even the very nearest stars are at least several light years away (see figure 4.1). The method of parallax, though, was rapidly found to be limited. Accurate twentieth-century measuring techniques were quite precise out to about 50 light years and usable to about three times that far. Unfortunately, by 500 to 600

Cosmological Enigmas

light years distance the parallax was so tiny that even the best techniques available in the 1970s could do little to measure it.

Late in the eighteenth century, Sir William Herschel gave a hint of how much bigger the Galaxy was than anyone had suspected. When Herschel discovered the planet Uranus, he was carrying out a "survey of the heavens." This had nothing to do with discovering new planets; Herschel wanted to map the distribution of stars in the sky. He assumed that the brighter a star is, the closer it is to us. Usually people describe this assumption as being a major error on Herschel's part. In reality, it is a gross approximation, but it holds a substantial element of truth. We now know that stars are of different sizes and different luminosities, with the most luminous stars being millions of times brighter than the dimmest. If we take stars of about the same luminosity, however, their relative brightness will give us a pretty good guide as to their distance. Of course, a faint star may be a luminous star at a great distance, or a dim star close by, but in general, farther is fainter.

Thus Herschel scanned the sky with his telescope, counting the stars visible in every region, and rapidly came up with various important truths. First, he established that the stars are not evenly distributed around the sky; there are many more toward the Milky Way, the dim band of light circling the sky that city dwellers no longer see. Second, he discovered that there are vastly more faint stars than bright ones and that, as we go to fainter and fainter magnitudes, the number of stars increases ever faster.

Herschel thus produced the first reasonably accurate map of our Galaxy, without even knowing that there was such a thing. Because many stars were concentrated in a thin band in the sky, they have to be distributed in a thin flat disk with a considerable extension, and our Sun has to be inside the disk. Herschel only committed one significant error: the vestiges of the Ptolemaic, Earth-centered model led him to place the Sun at the center of the Galaxy despite the fact that he saw many fewer stars on one side of the sky than on the other. This should have suggested to him that we are really well to one side. In fact, his final map of the Galaxy looked rather like a hamburger in which half the meat is missing, giving a big gap on one side between the halves of the bun![8] Although it would take well over a century for this error to be corrected, Herschel's map of the Galaxy was a big advance and hinted how distant most of the stars are. That its creator was an amateur astronomer with little formal education made it all the more amazing.

Herschel also looked at many nebulae in the sky using the 48-inch (1.2-meter) telescope that he had built himself and made an even more brilliant deduction, although it was largely ignored at the time: some of the nebulae were made of stars and were distant galaxies like our own. Herschel did not know it at the time, but with this insight he was close to multiplying the size of the Universe by a factor of a million.

Herschel knew intuitively that the Galaxy was immense, but as the parallax method works only for the closer stars, how could its size be measured?

The Cosmic Ladder and Pulsating Stars

To measure more distant stars, we must create a cosmic ladder that allows us to go further and further out, one step at a time. Although this exercise has dominated astronomy for more than a century, it remains controversial. Three basic techniques have been used: spectroscopic parallax, star clusters, and variable stars.

Spectroscopic Parallax

The first method assumes that stars of the same type are similar in their luminosity. If you measure the parallax of, for example, Sirius, you know its exact distance. Knowing its brightness, you know how luminous it is. You then find that another, more distant star has the same spectrum as Sirius and is thus of the same type. If you assume that it has the same luminosity as Sirius, you can then estimate its distance with some accuracy. This method is extremely powerful because the most luminous stars can be seen even when they are in other galaxies. If you can find, for example, 10 stars of the same type in a galaxy or a distant star cluster, you can average the measures and get a much more exact distance. A great advantage of this method is that it works with any kind of star, even if it is isolated.

Clusters

The clusters technique is an extension of the former method. Many stars live in family groups called clusters, which can be either an "open" or "galactic" cluster of stars that have formed together from the same cloud of dust and gas or a "globular" cluster of some millions of stars. There are many clusters around the

Cosmological Enigmas

sky, of which two of the most famous and easily seen are the Hyades and the Pleiades in Taurus. A cluster contains many stars of different types, which means that instead of using the brightness of just one star to estimate its distance, we can use dozens, or even hundreds at once. Fortunately, the Hyades cluster, which forms the face of Taurus (the Bull), is close enough that its distance can be measured by parallax, allowing the method to be calibrated.

Cepheid and RR Lyrae V

The best method available to astronomers uses the peculiar properties of "pulsating variables." These are stars that, as their name suggests, pulsate—they expand and contract like a heartbeat. The difference is that this "heartbeat" is fantastically regular, like an extraordinarily precise pacemaker. The first of these, the star Delta Cephei, is easily visible to the naked eye. It was discovered by David Goodricke as early as 1784 and beats every 5.37 days; it is this star that gives its name to the class of Cepheid variables. Later a second, less luminous class of stars called RR Lyrae stars was recognized. Named after RR Lyrae, the first to be recognized, they beat much faster, having periods of a few hours (see figure 4.2). In 1912 Henrietta Leavitt was studying variable stars in the Small Magellanic Cloud, one of the two bright irregular galaxies that orbit our Milky Way, when she noticed two things. First, all the RR Lyrae stars that she could find were about the same brightness. This led her to conclude, correctly, that they are all the same luminosity, which we now know to be about 90 times that of the Sun. Second, she noticed that the shorter the period of a Cepheid variable, the fainter it was. Because all the stars in the Small Magellanic Cloud had to be at almost exactly the same distance from the Sun, it was evident that by knowing the period of the variation one could calculate the luminosity of the star—it was only necessary to measure the exact distance to one Cepheid to calibrate the relationship. This we now know as the period-luminosity relationship, which is one of the bases of all modern astronomy.

There, though, lies the rub. The closest of all the Cepheid variables is the Pole Star—Polaris—itself. Unfortunately, Polaris is 680 light years away and well past the distance at which we can measure parallax. Delta Cephei is almost twice as far away at 1,300 light years and thus even worse.

Enter Harlow Shapley.

Between 1914 and 1921 Harlow Shapley, an American astronomer, studied globular clusters. He used the two biggest telescopes in the world at that time, first the famous 60-inch (1.5-meter) and then the new 100-inch (2.5-meter) Hooker reflector at Mount Wilson in California. He also worked on the problem of calibrating Henrietta Leavitt's period-luminosity relationship. By finding RR Lyrae and Cepheid variables in open clusters, he first measured the distance to the cluster and then calculated the luminosity of these stars. Once finished, he now had a method to find the distance to any galaxy or cluster that had Cepheid or RR Lyrae variables in it.

Shapley used this information to measure the size of our Galaxy. When he noticed that almost all the globular clusters are in a small part of the sky around the constellations of Scorpio and Sagittarius, he made the connection that William Herschel had not nearly 150 years before. Such a nonrandom distribution could not possibly be an accident. Shapley realized this could only mean that the clusters were surrounding the center of the Galaxy and could be seen from a considerable distance from the center. Herschel could have come to this conclusion himself, but the moment was not right for such a big step. One thing seemed certain: the Milky Way is about 100,000 light years across, and the Sun is some 32,000 light years from the center.

Several years were to pass before it was proved that many of the nebulae that astronomers could observe were really distant external star systems, most of them at many millions of light years. From the first measurement of the distance of the star 61 Cygni, however, the size of the Universe had suddenly expanded by another factor of 10,000.

Sizing It Up

One-cent coins can give us an idea of how far we have expanded our horizons. If we represent the distance from the Earth to the Sun—the Astronomical Unit— by a penny coin, Neptune, the most distant of the regular planets, will be just 30 cents or 60 centimeters away from the start point. The Oort Cloud of comets that surrounds our solar system and marks its limits will go out to about 2 light years away—that is 130,000 one-cent coins, or a $1,300 long line of pennies. We are still nowhere near the nearest star, which is the faint companion of Alpha Centauri known as Proxima Centauri. Proxima Centauri is $2,660 away (that is, a line of pennies 5.3 kilometers long). Betelgeuse, the giant red star in Orion, is a huge

Cosmological Enigmas

$330,000 away. If the line of pennies starts in New York, it will get as far as Washington, D.C., and almost back to New York.

Betelgeuse is a rather near neighbor in space. If we go further out and start exploring the Galaxy seriously, things start to get seriously expensive.

The edge of the Galaxy nearest to the Sun is an impressive $11 million away—that is, a line of pennies 23,000 kilometers long that now stretches more than halfway around the world from New York.

In Isaac Asimov's Foundation Trilogy, his future Galactic civilization is based on the planet Trantor, at the Galactic Center, and the heart of the civilization is in this central region where the stars are most thickly clustered. Trantor is $20 million away and our line of pennies now stretches around the Earth and returns to New York. The outer rim of the Galaxy, where the planet Terminus is situated, is a decidedly expensive $52 million away, and our penny line goes a full two and a half times around the world or a quarter of the way to the Moon.

But is it possible that we will one day sail the Galactic ocean and make port at another star?

Sailing the Ocean between the Stars?

In my library at home, I have dozens of books, probably more than 100, with stories based on interstellar travel not just being possible, but being cheap, easy, and an everyday occurrence. If we go to a (much smaller) part of my library, the books that deal with unidentified flying objects, or UFOs, we will discover there are many people who believe that these are an everyday occurrence and that the extraterrestrial equivalents of Captain Kirk visit us every day of the week.[9] Books like the Foundation Trilogy, Harry Turtledove's World War series, or the much older Lensman series of E. E. "Doc" Smith have given me hundreds of hours of pleasure over many years, and I enjoy immensely the various Star Trek series and Star Wars. We have to ask ourselves, though, how plausible it is that one day we will be able to navigate the spaceways.

Not everyone is optimistic. In 1986 Arthur C. Clarke published the novel The Songs of Distant Earth, in which he tries to imagine a future technology that would permit interstellar travel. In his introduction he comments that the novel was "directly—and negatively—inspired by the recent rash of space operas on TV and movie screens. . . . Even the very closest star systems will always be decades or centuries apart; no Warp Six will ever get you from one episode to another in time for next week's installment."[10] So, there are actually two issues that we must consider:

1. Is interstellar travel practical in any way? In other words, can we ever travel between the stars?
2. Can interstellar travel be carried out quickly so that the journeys will take weeks or months rather than centuries?

We have already launched our first interstellar probes, albeit unmanned. The Pioneer 10, Pioneer 11, and Voyager 1 and 2 probes are all heading out of the solar system and will eventually reach the stars. Although Voyager 1 was launched five and a half years later than Pioneer 10, on February 17, 1998, Voyager 1 overtook the Pioneer 10 probe and is now the most distant human-constructed object in space. As of March 18, 2005, Voyager 1 was 14,207,000,000 kilometers from the Sun and receding from it at 17.177 kilometers per second, having already traveled 16,576,000,000 kilometers along its curved trajectory since launch. Radio signals now take 13 hours, 7 minutes, and 48 seconds to reach the Earth from the probe.

Even at this velocity, how long will it take the probe to travel just one light year? The answer is 17,450 years. So, even this fastest of human space probes is not exactly designed for speedy interstellar journeys. Slowly but surely, though, both Voyager probes will reach the stars. In about 40,000 years, Voyager 1 will drift within 1.6 light years (15,000,000,000,000 kilometers) of a faint star, cataloged as AC+79 3888, in the constellation of Camelopardalis. In some 296,000 years, Voyager 2 will pass 4.3 light years (40,000,000,000,000,000 miles) from Sirius, the brightest star in the sky.

The chances that a little green man, or even a little green inhabitant of Sirius, will detect and intercept the probe are exceptionally remote, but, just in case, both Voyagers and both Pioneers are supplied with a plaque that will reveal where they came from. The Voyagers also contain a golden record of the sounds of Earth, with greetings in 55 languages. A diagram on the cover explains where the probe has come from and how to play the record; the package is even completed by a stylus thoughtfully carried on board the probe to play it (see figure 4.3).

To take 300,000 years to travel the distance to the sixth closest star system to the Sun does not make for convenient journeys. Captain Kirk would rapidly show his impatience with the engines if Scotty reported that rather than arriving in his habitual few hours, the journey would take some ten thousand generations and would not fit into the peak-time schedule.

Practical Journeys

There are at least two approaches to interstellar journeys. One is to contemplate continuous acceleration ships. Spacecraft, such as the Starship *Enterprise,* have the drive continuously functioning and, apart from other wizardry, can achieve tremendous velocities. There is an alternative: why hurry?

Some people have suggested "generation ships." They argue that rather than having a small ship that goes fast, you can just as easily—and perhaps more so— have a large ship that travels slowly. Make the ship as large as a small asteroid and give it a crew of thousands, maybe even millions. The ship would function like a planet, with food, water, and air being constantly recycled as they are on Earth.[11] By spinning the ship, gravity would be provided. Even if it took 10,000 years to reach its destination, generation after generation would be born on board and would allow the ship to carry on the mission. Such a generation ship would be an ideal colonization vehicle.

But a generation ship has some important drawbacks. No closed habitat, even a planet, can ever be 100 percent efficient (the great problem, for example, with Mars as a habitat, is its constant losses of atmosphere to space). For such a ship to work, the "degree of closing" (that is, the efficiency of recycling) must be extremely close to 100 percent, and the losses must be kept virtually to zero. Can a system be built that would remain habitable for long enough for the crew to survive the journey?

A second problem is more philosophical. How would the crew react to a journey of perhaps 50,000 years? Can we be sure that the 1,000th generation of the crew would share the aims of the first generation? Is there any possibility that the crew would even *remember* its mission, let alone be determined to carry it out? Various science fiction stories have explored this theme with consistent and depressing results. Whatever indoctrination the crew received, whatever high aims it set out with, its members would inevitably split into disparate groupings, perhaps with a ruling elite, who are the descendants of the original officers and engineers, and an underclass, giving rise to social instability and an inevitable conclusion. No human empire has lasted for even a tiny fraction of the time necessary to complete the journey, a fact that does not inspire us with confidence that any social grouping could remain stable for so many millennia.

In the novel *The Songs of Distant Earth*, Arthur C. Clarke proposes an alternative solution. Many stories have suggested using some form of hibernation to survive long space journeys. Such techniques are almost certain to be perfected in the foreseeable future, but their application is uncertain. It is possible to envisage a crew hibernating for months, or even years, but there seems to be no way of avoiding the progressive deterioration of the body, particularly the brain, and loss of information over long periods of time. The alternative suggested by Clarke is to take embryos frozen in liquid nitrogen. These would be unfrozen and fertilized on reaching their destination. Automated programs would then raise and care for the infants until they are adults. Such a system allows a first generation to be raised. Successive generations would then come about in the more traditional manner. Such a system would allow large groups of colonists to make journeys lasting hundreds, possibly thousands of years. Gradual damage to the DNA in the embryos by cosmic radiation would set a limit to the journey times possible using this technique. For longer journeys, one would have to contemplate some kind of system in which genetic information is stored in a memory bank and reconstructed at destination. By having various backups of the information that can be constantly checked against the primary data for damage, there is potentially no limit to how long it can be stored.

What about journeys at far greater velocities approaching those of light? Many science fiction novels contemplate ships that can accelerate constantly for months or years. One of my favorites is Ortega's torch ships, which ply the routes between the planets and eventually the stars in the Robert Heinlein novels *Farmer in the Sky* and *Time for the Stars*.[12] These rely on the direct conversion of mass into energy to propel them, permitting constant acceleration for years at close to Earth's gravity (thus avoiding the medical problems associated with long periods of low gravity). Physics, though, does not at present seem to allow this possibility of direct conversion of mass to energy. Even the matter-antimatter engines of the Starship *Enterprise* involve some huge problems of physics.[13]

Accelerating at a constant one gravity so that the crew of the ship would feel exactly the same weight as on Earth, the ship after just one day would reach an astonishing 850 kilometers per second. After one month it would be at nearly 10 percent of the speed of light, and relativistic effects would start to kick in. After a year you would now be traveling virtually *at* the speed of light. The most famous of the relativistic effects is *time dilation*—the slowing down of time. For an astronaut traveling close to the speed of light, time slows down in such a way that a journey that takes hundreds of years from the point of view of the Earth (launch a spaceship to Betelgeuse in the year 2,250, expect it back some time around the year 3,295) may only last 10 years from the point of view of the crew, who would age just 10 years while more than 1,000 years will have passed on Earth. Incredibly, with a ship accelerating at a constant gravity and the benefit of time dilation, some estimate that any part of the known Universe could be reached in 42 years of ship time! This allows our intrepid astronauts to make a journey of millions of light years in what, to them at least, is a reasonable amount of time. There would be undoubted difficulties in adjustment—even grave psychological problems—in knowing that possibly 100 times as much time will have passed on Earth and that you would come back to a completely changed planet with everyone you ever knew dead many centuries before. Imagine lifting a Norman knight from eleventh-century England and suddenly transporting him to twenty-first-century London; what are the chances that he would go insane from culture shock? Sadly they are rather large.

Less well known is another, particularly unpleasant side effect of space travel at close to the speed of light. Space is not quite empty. Each cubic centimeter of space contains approximately one hydrogen molecule. That does not sound like much, but close to the speed of light our ship turns into a giant atom smasher. Some 30,000,000,000 hydrogen atoms will crash into every square centimeter of the front of the ship each second, not to mention small grains of dust mixed in with the hydrogen. The result would be a lethal blizzard of radiation that would penetrate the hull of the ship and rapidly fry its occupants. The dust would sandblast the hull violently (a tiny dust grain would have the energy of a cannon shell). Our crew would most likely be dead long before getting close to the speed of light.

The suggestion has thus been made that interstellar journeys would have to limit themselves to a maximum of 10 percent of the speed of light. A journey to Alpha Centauri, the nearest star system to our Sun, and back would take close to a century to complete. Without the benefits of time dilation such a journey could only be undertaken using advanced hibernation techniques or with a generation ship. *The Songs of Distant Earth* contemplates that such a ship would still need a giant shield in front (Arthur C. Clarke suggests a huge iceberg) to protect it from collisions with interstellar dust. All this, though, seems possibly unduly pessimistic; one foreseeable advance in space technology is the use of some kind of electrical or magnetic shielding to deflect the dust and gas before it reaches the hull of the ship. Also, rather than giant irregular-shaped behemoths navigating the spaceways, interstellar ships that travel close to the speed of light need to be quite streamlined to cut through the gas and dust.

Some authors have suggested putting this hydrogen to good use. In 1960 R. W. Bussard at Los Alamos Laboratory proposed the interstellar ramjet. This is a design to capture the thin interstellar gas with huge magnetic collectors and channel it into a fusion motor. Like ramjet motors on aircraft that function in the thin high atmosphere, once a certain minimum velocity was attained, the gas would be sufficiently compressed to provide a constant fuel supply and constant thrust. This idea is extremely attractive as a 1,000-tonne spacecraft could be made almost all payload, with no need to store fuel on board. A ramjet in interstellar space would require a collecting funnel 2,500 kilometers across, but that does not seem an excessive extrapolation of future technology. The Bussard ramjet seems to be the most plausible technological solution to spaceflight that we can imagine at present, although not all scientists agree; some studies suggest that far from being an efficient driving system for an interstellar spacecraft, it would actually serve as an extremely efficient brake to slow the spacecraft down.

Wormholes and Warp Drives

But there are more exotic possibilities. "What if . . . " is a force that has driven not just science fiction but science itself for centuries. What if the speed of light is not the limit? What if it can be evaded? What if we can take a shortcut through space with a warp drive or using a wormhole? What seems certain, at present, is that the speed of light is a genuine physical limit. Many people misunderstand this. Einstein does not say that it is impossible to travel faster than light; he just says that it is impossible to travel *at the speed of light.* The speed of light is like a giant unclimbable wall that lies in front of our spacecraft. On the other side lies an infinite horizon with unlimited velocity but, if we try to climb the wall, the higher we reach, the higher it gets. The same problem exists getting back to "our" side of the speed of light barrier: when we try to slow down at our destination, we crash, once again, into the barrier of the speed of light. As it seems impossible

Cosmological Enigmas

to pass from *slower* than the speed of light to *faster* than the speed of light without passing *through* the speed of light, we must reluctantly reject this possibility. Even if the famed tachyons, particles that can only travel faster than light, do exist, it seems impossible to harness them from our side of the wall.

Much science fiction, even from authors of the importance of Carl Sagan, suggests using black holes—wormholes in space—to travel great distances rapidly, entering a black hole in one place and leaving it through another perhaps thousands or millions of light years away. In his novel *Contact*, Carl Sagan's protagonist traveled to the center of the Galaxy through a series of wormholes, and then in the film of the same name Jodie Foster made this selfsame journey. Wormholes cause huge divergences of opinion. In an author's note at the end of the novel, Sagan thanks Caltech professor Kip Thorne (a distinguished physicist) for generating the 50 lines of equations that described the space transport system used in *Contact*. Other scientists differ sharply and argue that travel through a black hole would not be survivable, would not offer rapid journey times, or would be uncontrollable. In a story published on May 23, 2005, entitled "Travel Ban—Do TV Time Travelers Need to Find Another Way to Get Around?"[14] BBC science reporter Paul Rincon summarized the results of two recent research papers that suggest such travel is not practical. When distinguished experts differ so strongly on whether something is possible, you can be sure that the whole issue is not well understood.

Finally, what of the Starship *Enterprise*'s warp drive? Curiously, a paper published in 1999, in the journal *General Relativity and Quantum Cosmology*, has suggested that such a thing could one day be possible. In 1994 the Mexican physicist Miguel Alcubierre had suggested that a space warp bubble could theoretically be made, although it would require more energy than exists in the whole Universe to make it. In 1999, though, Chris Van Den Broeck of the Catholic University in Leuven, Belgium, published a new analysis that reduced the amount of energy required by a small matter of 62 zeros.[15] These two studies do not say that it can or will one day be built, only that it is, theoretically, possible to do it. Until such time as it becomes possible, if it ever does, all interstellar travel will have to be done the hard way.

DISTANCES BETWEEN THE STARS ARE HUGE, so huge that we cannot easily imagine them. This makes interstellar travel far more complex and difficult than hopping into Luke Skywalker's speedster and arriving at another solar system in just a few hours.[16] However much we may dislike it, we are limited to the laws of physics, and they state that, even allowing for huge extrapolations of our technological capabilities, interstellar travel is unlikely to be quick, cheap, or easy. That does not mean that it is impossible, especially if we are patient and accept slow journeys, but it does make a regular interstellar mass-transportation system seem exceedingly unlikely. (With 100,000 million stars in the Galaxy and our Sun in an undistinguished outer part and so many more interesting places to visit, is it really likely, as the UFO buffs claim, that hundreds, if not thousands of interstellar visitors have come to Earth in the past few decades?) There also remains the rather large question of whether there is anyone or anything to find out there when we do eventually reach the stars. That is the subject of a later chapter.

SUGGESTIONS FOR FURTHER READING

Popular Books

Patrick Moore, *History of Astronomy* (London: McDonald, 1983).

> *This is a favorite book of mine, originally published as Astronomy in 1961. My own 1983 copy was the sixth revised edition. A wonderful, lavishly illustrated, classic review of the history of astronomy. Organized as 40 short chapters (most are only four to five pages of largish print and some are much shorter) with titles such as "The Story of Tycho Brahe" and "The King's Astronomer," this book is a wonderful introduction to astronomy from the earliest Chinese observers to the modern day.*

Science Fiction

Arthur C. Clarke, *The Songs of Distant Earth* (London: Grafton Books, 1986).

> *A novel by a master of science fiction who attempts to describe interstellar travel using realistic and plausible future technology. An entertaining and informative read.*

Robert Heinlein, *Time for the Stars* (London and Sydney: Pan Books, 1956).

> *A novel about future exploration of nearby stars using huge spaceships with a crew of 200. Much of the novel is remarkably plausible even 50 years later, despite it having been written some two years before Sputnik 1 was launched.*

More Advanced Reading

R. W. Bussard, "Galactic Matter and Interstellar Flight," *Astronautica Acta* 6 (1960): 179–94.

> *The original paper that proposes the use of the Bussard ramjet for interstellar flight.*

C. Sagan, "Direct Contact among Galactic Civilizations by Relativistic Interstellar Space-flight," *Planetary and Space Science 11* (1963): 485–98

> *A paper by Carl Sagan speculating on the practical methods of interstellar travel that might be used by other intelligent civilizations in the Galaxy.*

On the Internet

Greek Astronomy
www-groups.dcs.st-and.ac.uk/~history/HistTopics/Greek_astronomy.html

> *A wonderful historical compilation of Greek astronomy. Detailed, but easily readable, it is well referenced. You can click on any of the dozens of names mentioned in the text and find a detailed biography and synopsis of their work.*

Period-Luminosity Relation for Variable Stars
www.astronomynotes.com/ismnotes/s5.htm

> *A nicely mounted page that contains a number of simple animations that explain how astronomers measure the distances to distant stars and galaxies.*

Measuring the Distance to a Globular Cluster
www.astro.washington.edu/labs/clearinghouse/labs/DistM4/m4.html

> *This lab exercise allows you to reproduce one of the classical experiments in astronomy and measure the distance to the globular cluster M4 in Scorpio by measuring the light curve of a RR Lyrae variable in the cluster. The experiment is simple and well explained and requires no special apparatus.*

CHAPTER 5

How Old Is the Universe?

The age of the Universe is controversial. Historically, it has caused religion to confront science. If we believe the version in Genesis, the Universe was created in six days and its creation finished with the appearance of the first man and woman on Earth.

In 1664, near the end of a long and prodigious career as a religious scholar, Archbishop James Ussher announced that the world began on October 22, 4004 B.C. Archbishop Ussher was born in Dublin in 1581 to a wealthy Anglo-Irish family. As a boy, he exhibited a gift for languages and entered the newly founded university Trinity College, Dublin, when he was just 13 years old. Ussher graduated at 19, received his master's degree at 20, and was professor of theology at the university at the exceptionally young age of 26. From 1625 to 1656 he was archbishop of Armagh and (Protestant) primate of All Ireland. In his most famous work, the *Annales veteris testamenti, a prima mundi origine deducti* (Annals of the Old Testament, deduced from the first origins of the world), published in 1650, he added the ages of the patriarchs given in Genesis 5 and concluded that the Earth was created on the evening preceding October 23, 4004 B.C., although other works claim that the date was October 26 (which actually corresponds to the date of the creation of Adam).

Archbishop Ussher's work was regarded as so authoritative that nobody questioned it for at least a century. Even today, creationists take his work to be definitive. By the end of the nineteenth century, however, evidence was growing that the Earth must be very much more than 6,000 years old.

Resisting Archbishop Ussher

Although there was no "eureka" moment when scientists had proof positive to refute Archbishop Ussher, his conclusions became progressively less and less tenable. The discovery of fossils of creatures that were not of any creature known on Earth could be dismissed as the work of the devil who had put them there to confound us. But, as more and more fossils were discovered of ever more different types and families, geologists became convinced they were seeing ancient forms of life that had died and been turned into rock. Knowing how slowly rocks formed it was obvious that they had to be far more ancient than 6,000 years old, but how much older it was impossible to say.

Geologists and physicists followed three lines of evidence that made plain that the Earth had to be much older than Archbishop Ussher supposed.

The Saltiness of the Oceans

Unlike lakes, oceans are salty. Salts and minerals in rocks are dissolved in rivers and the salt is carried down to the oceans. River water, though, is fresh with very little salt in it. Water in the oceans evaporates slowly, though, while the dissolved salt remains. So, over time, the water gets more and more concentrated. This process is extremely slow because the volume of water in the oceans is vast and the amount of water that evaporates each year is just a tiny fraction of the total.

By 1899 calculations showed that if the oceans started out as fresh water, it would take about 100 million years for them to become as salty as they are now.[1]

The Thickness of Sediments

In the late nineteenth century geologists began to understand rock formation. The process is similar to the build-up of salt in the oceans. Geologists knew that, quite apart from carrying fresh water and dissolved minerals to the oceans, rivers also carry large quantities of silt. Geologists could see formations in many places that were obviously made from silts that had been compressed until they were turned into rocks. Knowing at what rate silt layers built up in river mouths and assuming that this rate was the same in the past, it was possible to calculate how long was required to build up the layers of sedimentary rocks from them.

The result was that about 100 million years would be needed. This result, pleasingly, was identical to the estimate of the age of the Earth that came from calculating the build-up of salt. Even had it not been for the third line of evidence geologists could have been quite certain that the Earth was 100 million years old.

We now know that the Earth is far older than this. Why did the geologists get it wrong? The answer is exactly the same reason why the age of the oceans was "wrong" (it just did not correspond to the real age of the Earth). It was not until the 1960s and 1970s that geologists realized that continental drift was a fact of life and accepted it completely. As early as 1912 the similarity in the shape of South America and Africa and the similarities of the flora and fauna in the two continents led Alfred Wegener to suggest that they had once been joined and to suggest the theory of continental drift. However, the difficulty of imagining just how continents could move and drift around made the theory difficult to accept, and by 1928 they had been rejected by the scientific community. In Europe, the theory started to be accepted in the 1950s and in North America in the 1960s, when finally an explanation of how continental drift could happen was proposed.

Continental drift showed that rocks, continents, and oceans were constantly being born and dying. The age of the sediments that geologists found around European coastlines came out at about 100 million years because that was the age that the Atlantic sediments had as the Atlantic Ocean started to open up about 100 million years ago.

Curiously though, a third line of evidence also gave this correct, but misleading, result.

The Rate of Cooling of the Earth

Knowing that the interior of the Earth is hot, as is shown by volcanoes and deep mines, the great physicist Lord Kelvin in 1870 calculated how long it would take the Earth to cool to its present temperature if it had started molten. He assumed that the temperature increased by 30°C for each kilometer of depth (this

figure is a little too high—the true value in the upper crust of the Earth is 25°C per kilometer of depth). Calculating the rate over cooling with time, he found that the Earth had to be between 30 and 100 million years—the former value if the Earth had cooled faster in the past when it was hotter, the latter if the rate had been constant over time.

Once again the figure of 100 million years for the age of the Earth appears, and once again it is misleading. In this case, the reason remained a mystery for a shorter time, although the result caused massive dissension. The answer to the mystery was, of course, radioactivity. Although the Earth is cooling—heat is lost from the interior through volcanoes, geothermal vents, and the crust itself—the heat is being constantly replenished inside. The decay of radioactive elements, particularly uranium, thorium, and potassium, supplies almost enough heat to replace what is lost.[2] This heat is such that the real age of the Earth is actually about 150 times greater than Lord Kelvin had estimated.

The Age of the Earth

Once scientists became aware of radioactivity, they had an immensely powerful tool potentially available to calculate the ages of rocks, be they on the Moon, on other planets, or in a meteorite. Radioactive decay was discovered by accident in 1896. Antoine-Henri Becquerel stored some photographic plates with some pieces of uranium-bearing minerals. Developing the plates later, he found they were darkened where the minerals had rested on them. He realized that the minerals emitted something, evidently some type of radiation, that affected the emulsion of the photographic plates in the same way that light did. Two years later, in 1898, Marie Curie who had emigrated from her native Poland to Paris so that she could study science, followed up this discovery by studying thorium-bearing minerals for her Ph.D. She studied the characteristics of the radiation emitted by the minerals and reached a series of fundamental conclusions that were, in 1903, to win her the first ever Nobel Prize for Physics.[3] First, the amount of radiation was directly proportional to the amount of thorium in the minerals. Second, the amount of radiation did not depend on the pressure, or the temperature, or any other physical property of the minerals that she could vary. This made it totally unlike any known chemical reaction. Later it was demonstrated that the radiation could not be created or destroyed by any physical or chemical treatment of the minerals.

Ernest Rutherford, born in New Zealand, the son of a Scottish émigré wheel-wright, was also to become a Nobel Prize–winning physicist. In 1901 Rutherford suggested that one element changes to another in radioactivity, thus establishing the idea of radioactive decay. Soon it was demonstrated that particular elements would decay at a constant rate into other elements, and that the relative amount of each of the two elements in a particular sample of rock could be used to calculate its age (see figure 5.1).

The oldest rocks found on the Earth, zircon crystals in a block of granite in Western Australia, have been dated at 3,960 million years old. Rocks almost as old (3,800 million years old) have been discovered in a number of places around the world, including Greenland, Antarctica, South Africa, Minnesota, and Wyoming. The Earth's continents are unstable, always being renewed and destroyed by continental drift, which means that old rocks are unlikely to survive. Thus, rocks from the very earliest days of the Earth are almost certainly destroyed. Fortunately, scientists still have access to much older rocks.

When astronauts went to the Moon, one of their aims, at least in the later missions that were dedicated to science and geology rather than public relations, was to find the "genesis rock"—a rock from the original lunar bedrock that was not later melted and modified. The Apollo 15 astronauts thought they had succeeded, but the rock was later found to be more recent, although still as ancient as the oldest rocks on Earth. Subsequently, rocks as old as 4,400 million years old—sometimes written as 4.4 Gyr, or 4.4 gigayears—have been found in the lunar samples returned to Earth. Even so, these are not the oldest rocks available to us. Meteorites have been found with an age of 4.6 Gyr. These are the leftovers from the formation of the solar system itself and are the oldest objects in the solar system.[4] Most were formed in small asteroids that cooled rapidly, so the rocks in meteorites are almost as old as the solar system itself. The age of the Earth is generally estimated to be about 4.7 Gyr.

We know, however, that both the Earth and the Sun are much younger than the Universe.

The Elements and the Stars

Stellar births and deaths give us information about the ages of the stars and of the Universe. In the Big Bang, which we will meet intimately later, the only ele-

Cosmological Enigmas

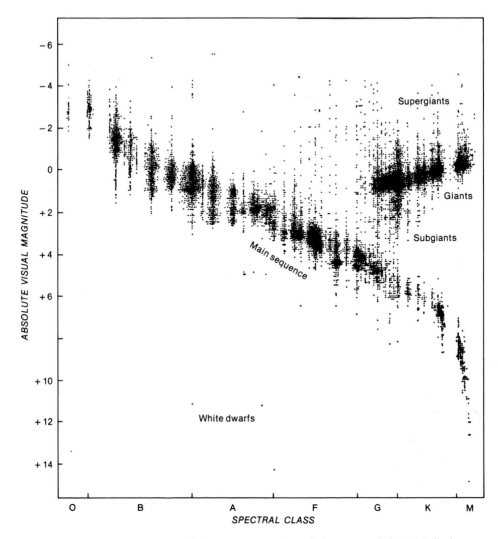

Figure 1.1. An early representation of the Hertzsprung-Russell diagram made by W. Gyllenberg at Lund Observatory (Sweden). The diagonal band crossing the diagram is the Main Sequence. Note how the luminosity of the very coolest red stars drops off rapidly in the bottom right-hand corner of the diagram.

Figure 1.2. The Orion Nebula, a stellar birthplace.

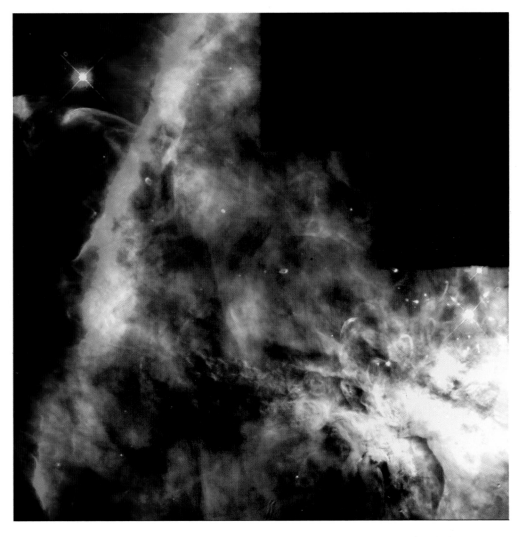

Figure 1.3. Zooming in on star formation. The Hubble Space Telescope panorama of turbulent star-forming regions in the center of the Orion Nebula. In this region of the nebula stars are being formed by the collapse of the gas cloud. Note the oval shape in the center of the image. This is a star in process of formation, with a disk around it that may one day condense into planets.

Figure 1.4. The Veil Nebula, an interstellar shock front. This photograph shows just a small part of what is a large ring of gas and dust, around three degrees across, in the constellation of Cygnus. It is the remnant of a supernova explosion that took place some 5,000 years ago. Estimates suggest that, at its distance of approximately 1,400 light years, this supernova would have reached magnitude 8. The leading edge is blue where it is still impacting at high velocity with the interstellar gas and exciting it.

Figure 1.5. The so-called Pillars of Creation, a small part of the nebula Messier 16, the Eagle Nebula. This beautiful image by the Hubble Space Telescope shows star formation in action and an alternative process for generating the collapse of a cloud of dust and gas. In this nebula, we see three pillars that are similar to the famous pillars that form in desert regions when a stone protects the soft rock below from erosion from occasional intense rainfall. In this case there is a dense cloud called an EGG (evaporating gas globule) inside the peak of the pillar, which protects the column from "erosion" from the intense stellar wind—the tenuous gas blown off at high velocity like the solar wind that causes the tails of comets—by highly luminous young stars just out of the field of view. The impact of this intense stellar wind in the gas cloud causes it to start to collapse.

Figure 1.6. Praesepe. We see a relatively old cluster, Messier 44, Praesepe, or the Beehive, in Cancer. Although only 400 million years old, that is, a small fraction of the age of the Sun, this cluster is old enough for all traces of gas and dust to have disappeared, although the stars have not yet separated and split up, as is usually the case in old clusters of this type, and are still all contained in a diameter of just 10 light years.

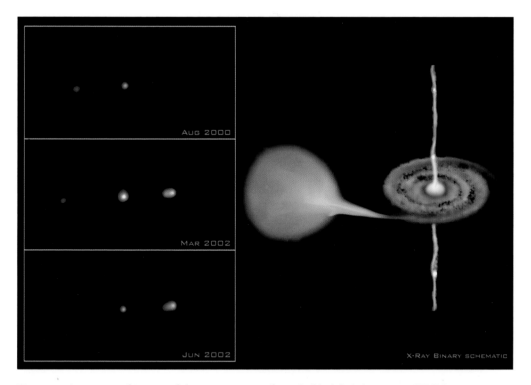

Figure 2.1. A montage of images of the x-ray emission from the black hole binary star XTE J1550-564 taken between August 2000 and June 2002 with the orbiting Chandra X-ray Observatory. The three images on the left show the star (centered) with a jet of x-ray emission bursting away first on the left-hand side and then on the right. On the right is an artist's concept of how this binary system consisting of a red giant star and a black hole might look if we could see it from close up. Material from the red giant star spirals onto the black hole, while the excess that the black hole cannot absorb squirts away at the two poles.

Figure 2.2. The center of the galaxy M87 in Virgo. A long, straight jet leaves the nucleus of the galaxy for hundreds of thousands of light years. The only object that we can think of that can manage to keep such a persistently straight orientation over such a long time is a very massive gyroscope, like a giant black hole. The beam of gas only starts to spread out thousands of light years from the center as the gas and dust that it hits inside the galaxy begin to disperse it.

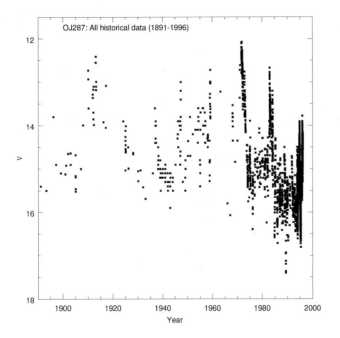

Figure 3.1. The light curve of the blazar OJ287 between 1893 and 1996. We can see how there seem to be regularly spaced, sharp maxima throughout the light curve. These are the outbursts that occur every 11.5 to 12 years when the two black holes are at their closest approach in their orbit. A new outburst duly occurred in 1994–95, as predicted.

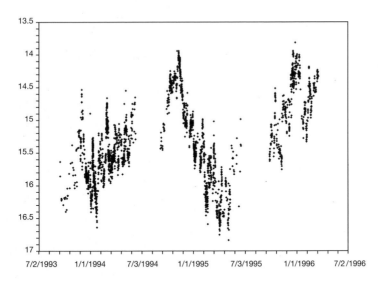

Figure 3.2. The light curve of OJ287 between 1993 and 1995. Two big brightenings are seen that seem to confirm the predictions of the binary black hole model.

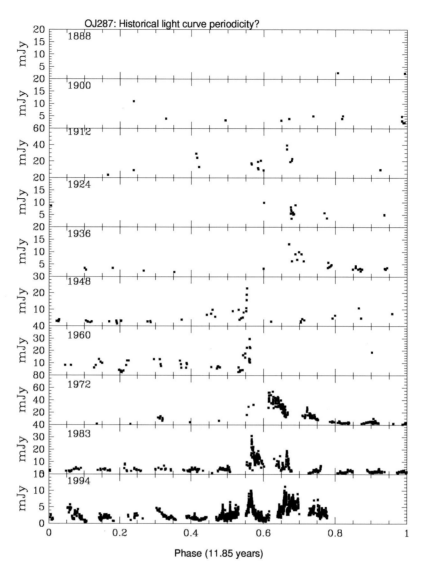

Figure 3.3. The same light curve as in figure 3.1 but with the data lined up under the assumption that there is a period of 11.85 years in the outbursts. We can see how the big maxima line up, although there is no information on them prior to 1948, as there are too few observations to cover the light curve well enough before then. The vertical scale is in what astronomers call flux units rather than magnitudes; the advantage of these is that, unlike magnitudes, twice as big does genuinely mean twice as bright.

Figure 3.4. The original ROTSE (Robotic Optical Transient Search Experiment) telescope photographed with Jim Wren and Robert Kehoe, two of its operators in 1998. The telescope consists of four 200-millimeter telefocal lenses attached to CCD cameras and on a computer-controlled mount.

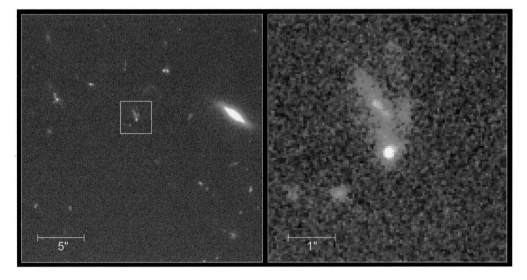

Figure 3.5. A Hubble Space Telescope image of the faint afterglow of the gamma ray burst GRB990123. The right frame shows a zoom on the GRB afterglow (the bright point) and its associated galaxy. It (the faint smudge) appears to be in two galaxies that are colliding or very close to each other. Such galaxies are often associated with huge amounts of star formation that involved the formation of many massive stars.

Figure 3.6. The Pistol Star (the bright star in the center of the image) and the Pistol Nebula, which gives the star its name. This star, imaged by the NICMOS infrared camera of the Hubble Space Telescope, appears to be the most massive and luminous star in the Galaxy. This star is a candidate to turn itself into a hypernova and thus a gamma ray burst.

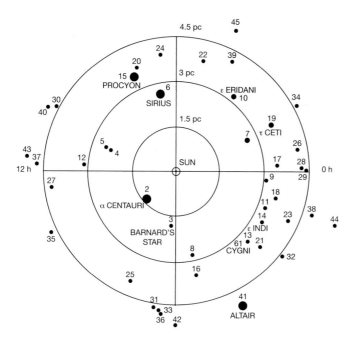

Figure 4.1. The distribution of the stars closer than 5 parsecs (16.3 light years) from the sun. The more luminous the star, the bigger the point. The number by each star is the order of distance from the sun.

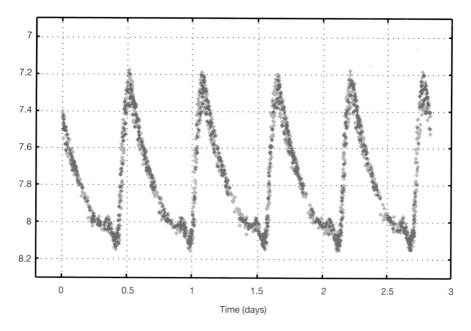

Figure 4.2. The light curve of the star RR Lyrae, which is the prototype of its class of pulsating variables. It cannot be seen with the naked eye but is quite easy to see in binoculars. Note how it changes brightness by almost a magnitude in just over 12 hours. Some RR Lyrae stars change brightness so fast that the changes are obvious over a few minutes. What makes them special for astronomers is the fact that all are almost exactly 90 times as luminous as the Sun: find an RR Lyrae star, and measure its brightness and you know its distance.

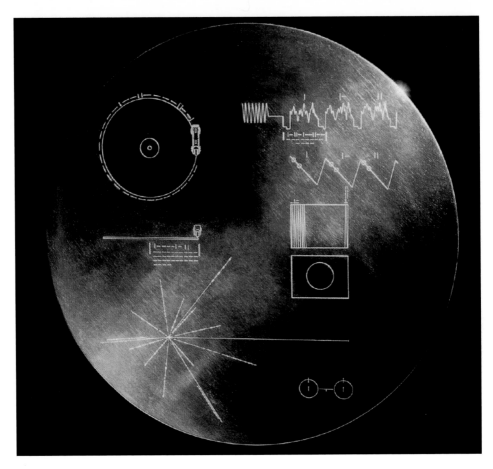

Figure 4.3. The golden record carried by the two Voyager probes, which has greetings in 55 languages, 115 images, and a variety of sounds and music from Earth, along with the translation of the symbols and images appearing on the cover that are intended to let any civilization that encounters the probe play the record.

Geologically Useful Decay Schemes

Parent	Daughter	Half-life (years)
^{235}U	^{207}Pb	4.5×10^9
^{238}U	^{206}Pb	0.71×10^9
^{40}K	^{40}Ar	1.25×10^9
^{87}Rb	^{87}Sr	47×10^9
^{14}C	^{14}N	5730

Figure 5.1. A table of some of the radioactive decays that geologists find useful to estimate the ages of rocks and that have helped to establish the age of the Earth. One of the most important is the decay of rubidium to strontium. In rubidium decay, a neutron in the nucleus turns into a proton and releases an electron. The half-life of this process, that is, the time taken for half of the rubidium to turn into strontium, is 47,000 million years; this makes it ideal to date very ancient rocks in which other radioactive elements have largely decayed away.

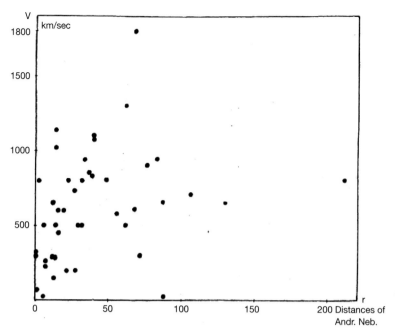

Figure 5.2. Knut Lundmark's 1924 plot of the velocity of recession of galaxies against their distance. Had it not been for a single bad point in the plot, he would almost certainly have beaten Edwin Hubble to discover the expansion of the Universe and "Hubble's Law"—which states that more distant galaxies recede more rapidly and which allows us to measure the distance of the furthest objects in the Universe from their red shift—and quite possibly we would be discussing the Slipher-Lundmark Law.

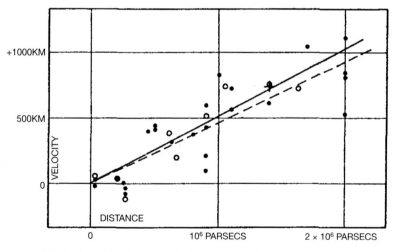

Figure 5.3. Hubble's plot of the distance of galaxies against their red shift, published in 1929. Even this rather low-quality plot is sufficient to show the difference with Lundmark's version of the same plot published five years earlier. With his access to a much bigger telescope than Slipher had available to him, Hubble could measure much more distant galaxies and also obtain better accuracy, hence the clarity of his plot. The slope of the fitted line gives the Hubble constant.

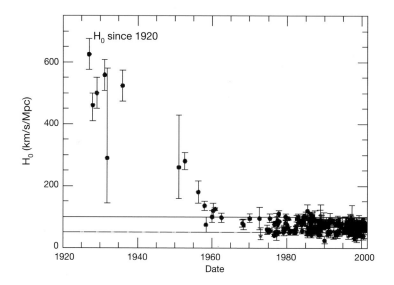

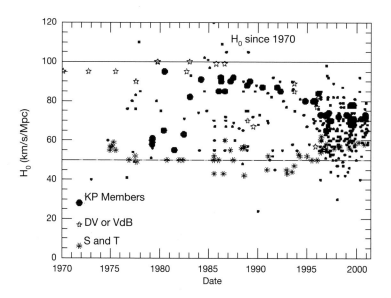

Figure 5.4. Plots prepared by John Huchra at the Harvard-Smithsonian Center for Astrophysics showing all the different values of the Hubble constant measured by different groups over the years. In the top panel we see how the first values were rapidly revised downwards until they reached a steady, constant, but rather broad range from the 1970s onwards. In the bottom panel the values measured since 1970 are plotted on an expanded scale. Values by different groups are identified: stars for Sandage and Tammann; pentagonal stars for Sidney van den Bergh and Gerard de Vaucouleurs; large dots for astronomers who would later use the Hubble Space Telescope to estimate the Hubble constant; and small dots for everyone else. The values lie over a range of a factor of about 5 from 25 to 125 km/s/Mpc.

ments that were formed were hydrogen and helium— approximately 77 percent hydrogen and 23 percent helium, nothing else. Yet we know that stars such as the Sun contain approximately 1 percent of heavier elements (astronomers lump all these together as "metals," although most are not metals). When we look at two apparently identical stars, we often find that the spectrum of one shows much stronger metal lines than the other. In the argot of the astronomers—and, again, it is a total misnomer—we say that one star is more metallic than the other. The reason is that the first stars that were formed in the Universe were made of original Big Bang material with only hydrogen and helium. When these stars died, they seeded the interstellar medium with the first heavy elements. These were incorporated into the next generation of stars that were born, which, in turn, died and seeded the interstellar medium with even more heavy elements. Over thousands of millions of years, the amount of heavy elements in the Universe increased gradually. Thus, the amount of heavy elements (or "metals") that we see in the spectrum of a star tells us how old that star is.

Metal-poor stars are stars made from very old material that had not yet been enriched with heavy elements. We know approximately how fast the interstellar medium is being enriched with heavy elements, so just by measuring the amount of metals in a star, we can estimate its age; a very metal-poor star is an extremely old one, and the lower the amount of metals, the older it is. The oldest stars are those in the globular clusters that orbit galaxies, including our own, and that were formed from the purest, most uncontaminated gas from the Big Bang.

To estimate the ages of these stars accurately, we have to use a clever statistical trick. We know that the more massive a star, the more rapidly it uses its hydrogen fuel and the shorter it lives. The Sun will continue to "burn" hydrogen in its nucleus for 10,000 million years before the crisis occurs and the outer layers swell as it turns into a red giant. We know then that, if we see a cluster of stars in which stars like the Sun have turned into red giants, that cluster must be at least 10,000 million years old. If even less massive stars than the Sun have become red giants, then they are older still. In contrast, if the cluster of stars is very young, only the bluest and most massive stars with their very short lives will have had time to exhaust their hydrogen. By looking at the exact point where the Main Sequence stops and the stars leave it, we can estimate with great exactitude the age of the cluster.

However, there is a catch.

To know how massive a star is, you have to know how luminous it is. And to know how luminous it is, you must know how far away it is. And there lies the rub.

Getting the Range

Globular clusters and, even more, galaxies are far too remote to use parallax. We rely basically on the distance-luminosity relation for Cepheid variables and RR Lyrae stars. The exact calibration of this relation and measurement of distances in the Universe is such a fundamental problem in astrophysics that that Hubble Space Telescope has dedicated a great deal of its observing time to the issue.

The problem looks simple: if we can see Cepheid and RR Lyrae variables, then we can calibrate their distance from the period-luminosity relationship. Unfortunately, things are never that easy. For a start, the luminosity of these stars depends to a small degree on their composition—the amount of heavy elements they contain.[5] Also, as we look at a star through the dust and gas of our Galaxy, or through the dust and gas of another galaxy, it seems dimmer than it really is because the material in the way absorbs some of the light. Thus we have to correct the apparent luminosity of the stars for these effects and estimate the real luminosity of the star. This is no simple matter, not to mention that one has to recalibrate the period-luminosity relationship as exactly as possible too.

In the end, a combination of observations by the Hubble Space Telescope of globular clusters and observations by a European satellite called Hipparcos have worked the trick. Hipparcos, which carried a 29-centimeter telescope, has been a remarkable success story despite almost being lost; its aim was to measure with high precision, over two and a half years, the positions and parallaxes of 100,000 stars, allowing their distances to be determined. The precision was such that stars up to about 1,500 light years away could have their distance measured accurately, about 10 times further away than we could do from the Earth's surface.

The satellite was launched from Kourou, French Guyana, into a geostationary transfer orbit on August 8, 1989, but it did not reach its intended geostationary orbit after its apogee boost motor failed (the motor should have fired when Hipparcos was farthest from Earth in its orbit to push it into a circular geostationary orbit). As a result, Hipparcos went into a highly eccentric orbit, and for a time it was feared that much, if not all, of its intended scientific results might be lost. Great work by the science and engineering teams at the European

Space Agency enabled Hipparcos to achieve and to surpass all the aims of the mission.

It was hoped that Hipparcos would measure 100,000 stars with high accuracy. In fact, it did even better. Some 120,000 stellar distances were measured with high precision until communication with the on-board computer was lost on August 15, 1993, and the mission was terminated. The results were interesting. Thanks to Hipparcos, astronomers could look statistically at the brightness of these same types of stars in globular clusters and so determine their distance far more accurately than ever before. The results, published in 1998,[6] suggested that the globular clusters around our Galaxy are farther away than had previously been thought. The stars in them, therefore, are more luminous than had been previously supposed—and therefore they are younger.

Our current best guess, using Hipparcos data, is that the oldest stars in globular clusters are 11.5 ± 1.3 Gyr old.[7] This is substantially less than the previous estimate of 14 Gyr, which was embarrassing to astronomers, as this was older than the estimated age of the entire Universe—more about that later. In the last few years, however, astronomers have recognized there are some stars that are significantly older than those in globular clusters. Our Galaxy is surrounded by a loose cloud of stars called the Galactic Halo. Some stars in this halo are extremely poor in metals and are thus extremely old, having formed from ancient, metal-poor gas early in the life of the Universe. The uncertainty in the ages of these stars is large indeed, but it is almost certain that they are older, possibly much older, than the stars in globular clusters.

By calculating the distances to the globular clusters more precisely, we can also calculate more accurately the period-luminosity relationship for Cepheid and RR Lyrae stars. This has led to a significant change in our ideas about the age of the Universe itself, for we can now determine the distances to other galaxies with much greater accuracy.

From Pluto to the Cosmos!

Back in the 1920s a small group of astronomers was working to restore credibility to Lowell Observatory in Flagstaff, Arizona, whose reputation had been severely tarnished.[8] The three remaining astronomers on the staff in particular faced this seemingly thankless task in an observatory starved of funding: Carl Otto

Lampland, Earl Carl Slipher, and his older brother Vesto Melvin Slipher. Of these, the largest burden fell on Vesto Slipher, born in Mulberry, Indiana, in 1875, who took over as the director of the observatory on Percival Lowell's death.

It is a scandal that Vesto Slipher has not been more widely recognized. Although he provided the impulse for Clyde Tombaugh's discovery of Pluto, his main interest was the study of galaxies. In 1914 Slipher gave a presentation to the seventeenth meeting of the American Astronomical Society entitled "Spectrographic Observations of Nebulae."[9] Slipher concentrated on the spiral nebulae that for many years had been suspected of being made of stars and which we now know as galaxies. He began observing the spectra of galaxies in 1912, using the 24-inch (60-centimeter) reflector at Lowell Observatory. He immediately made the startling discovery that the Andromeda nebula shows a substantial blue shift in its spectrum. This blue shift—the lines in the spectrum shifting toward shorter wavelengths rather than the red shift seen in most galaxies—is the Doppler effect of the galaxy approaching the Sun. Slipher showed that the Andromeda nebula was approaching us at about 300 kilometers per second. He then observed 14 other spiral galaxies whose spectra were made up of the light of many stars, the "average" of which was of a star of spectral type G or K (yellow to orange). Of the 14 additional galaxies, 11 showed a red shift, demonstrating that they were receding from us; three had red shifts of 1,000 to 1,100 kilometers per second.[10] In a remarkable show of admiration for his work, the assembled audience gave Slipher a standing ovation when he finished his presentation—a rare tribute in the astronomical world.

In April 1917 Slipher read a paper to the American Philosophical Society in which he increased to 25 the number of galaxies observed, finding that 21 showed a red shift and were thus receding. In this paper he makes a remarkable comment: "For us to have such motion and the stars not show it means that our whole stellar system moves and carries us with it. It has for a long time been suggested that the spiral nebulae are stellar systems seen at great distances. . . . This theory, it seems to me, gains favor in the present observations." Thus as early as 1917 Vesto Slipher was presenting imposing evidence that the galaxies are external systems, and he was very close to discovering the expansion of the Universe;[11] if he had had access to a telescope larger than the 60-centimeter reflector at Lowell, it is entirely possible that he would have made this major discovery. The credit, however, went elsewhere.

In 1924 a study of Slipher's results combined with other observations (and containing fulsome tribute to Slipher) by Knut Lundmark at the Royal Greenwich Observatory almost stumbled on the expansion of the Universe.[12] Lundmark was looking for correlations of properties such as distance and velocity for different types of objects, including stars in our Galaxy and spiral galaxies (the latter based mainly on Slipher's work). One of the plots Lundmark showed in his paper—reproduced in figure 5.2—was that of the velocity of galaxies against their distance, estimated in terms of the distance of the Andromeda Galaxy (he gave an excellent estimate of the real distance to the galaxy based on the brightness of novae, significantly better than the value that was accepted until 1950). If one of his galaxies (NGC 584) had not been badly placed on his plot (it was estimated to be further away than it really is), Lundmark would probably have discovered the expansion of the Universe, and then the law that allows us to calculate the distance to galaxies and quasars in the furthest reaches of the Universe would be known as the Slipher-Lundmark Law. That honor would go to Edwin Hubble.

Hubble's Law

In 1999 *Time* magazine elected its 100 most important and influential people of the twentieth century. It was no great surprise that Albert Einstein was nominated its "person of the century" ahead of FDR and Mahatma Gandhi. Less well known than such luminaries as Einstein, Salk, Freud, and Churchill was Edwin Hubble. The name "Hubble" has become familiar to millions over the years thanks to the Hubble Space Telescope, but how many people know who Edwin Hubble was? Born in Marshfield, Missouri, in 1889, Hubble moved with his family to Chicago in 1898, where he shone more as an athlete than academically. As a young man, he was even offered a professional contract as a boxer. Although he majored in science at the University of Chicago, he nearly turned away from science. The new graduate won a Rhodes scholarship to Oxford, where he studied law because of a deathbed request from his father.[13] On returning to the United States, he spent a year teaching high school Spanish. But then Hubble decided to take up a postgraduate position at Yerkes Observatory in Wisconsin, and, as they say, the rest is history.

Hubble's work was impressive enough for him to be offered a job at Mount Wilson Observatory, which housed the two largest telescopes in the world.[14] First, though, he volunteered for the army when the United States entered World War I in 1917. When he returned to Mount Wilson in 1919, it was as "Major Hubble, if you please."[15] It did not take long for a rivalry to start between Hubble and Harlow Shapley, the man who had earlier measured the size of the Galaxy. Shapley moved to Harvard in 1921, still believing that galaxies were not external star systems. Hubble was determined to prove him wrong (one gets the impression that there was a personality clash between them, with Hubble less than totally respectful of his senior's reputation).[16] Two years later, Hubble used the new 100-inch (2.5-meter) Hooker reflector at Mount Wilson and found a Cepheid variable in the Andromeda Galaxy. Applying Shapley's own technique against him, Hubble used the period-luminosity relationship to demonstrate that the Andromeda Galaxy was 900,000 light years away and thus well outside our own Galaxy, which Shapley had estimated to be 300,000 light years across.[17] Shortly afterward, Hubble repeated this feat by detecting Cepheids in the Triangulum Spiral, M33 (another of our closest neighbors in space), and then, with the galaxy NGC 6822, demonstrating that the Andromeda Galaxy was not an exception and that all the spiral nebulae were outside our own. The scene was set for Hubble not only to extend Slipher's work and Lundmark's graph but even to correct Albert Einstein's work.

Following his general theory of relativity, published in 1915, Albert Einstein came to an uncomfortable conclusion. His equations predicted that the Universe could not be static—it had to be either expanding or contracting. To correct this, he added to the equations the "Lambda term," or cosmological constant, to counteract gravity, to stabilize the Universe, and to avoid the need for expansion. Einstein later called this his greatest ever blunder;[18] it was Hubble who showed him that he was wrong and that the Universe was expanding. Aided by Milton Humason, Hubble set out to measure the red shifts and distances of galaxies using the 100-inch Hooker reflector at Mount Wilson.

By 1929 Hubble started publishing a series of earth-shattering results. His first paper showed what others had previously suspected: based on 24 galaxies for which he could measure the distance, there was a linear relationship between the distance of a galaxy and its red shift and thus velocity of recession.[19] This original plot is shown in figure 5.3. It is hard to believe that such an apparently low-quality graph could have such profound consequences for science. The plot clearly shows that the more distant the galaxy, the greater its red shift and hence its velocity of recession. Part of Hubble's success was that he realized that Slipher was right: there

Cosmological Enigmas

was an individual, or peculiar, solar velocity (we now know that this is about 300 kilometers per second approximately in the direction of Virgo), and, taking this effect into account, he was able to correct Slipher's own plot.

By 1930 Humason had obtained red shifts of galaxies that reached 20,000 kilometers per second—around 7 percent of the speed of light—and in 1936 Hubble and Humason published individually results for galaxies in Ursa Major showing that their recession velocity was a massive 40,000 kilometers per second (13 percent of the speed of light).

Hubble's first estimate was that the velocity of galaxies increased by 500 kilometers per second for each Megaparsec (million parsecs) of distance (usually written as 500 km/s/Mpc). The slope of this graph is called Hubble's constant.[20] Nothing can travel faster than the speed of light, so if a galaxy recedes with speed, it has to be at the edge of the visible universe. So Hubble's law gave an idea of how big the Universe is. But a galaxy at the edge of the Universe has taken the age of the Universe to get there! So the reciprocal of Hubble's constant is the age of the Universe. Hubble's initial plot gave 2 Gyr—200,000 million years—which assumes that the galaxies neither accelerate nor slow down at great distances. (This also assumes that the Universe is effectively empty and that its mass is small, so its force of gravity is insignificant and does not slow the recession of the galaxies appreciably.) At the time, this value seemed reasonable enough, although it would come rapidly into conflict with the age of the Earth derived from radioactive decay, which in 1936 was measured to be a little more than 3 Gyr. Such was Hubble's stature that no one questioned his results.

In 1949 the new 200-inch (5-meter) Hale reflector at Mount Palomar in California entered service. By now Hubble was too ill to spend his nights in the dome with the telescope. Walter Baade, ironically a collaborator of Hubble, started a program of observations with the new telescope that showed that Hubble had committed a serious error in his earlier work. Baade found that the Cepheids in the Andromeda Galaxy were not the normal Cepheids that Hubble had believed but were more luminous. At a stroke it was realized the Andromeda Galaxy was more than twice as far away as Hubble had believed—2.2 million light years rather than 900,000 (Lundmark's almost forgotten estimated distance in 1924 had been far closer than Hubble's). With Baade's work, the size of the Universe increased rapidly, and another curious anomaly in Hubble's work was explained. According to his distances, the Milky Way was one of the largest galaxies in the known

Universe, but Baade demonstrated that this proposition is far from true. The Milky Way is much smaller than, for example, the Andromeda Galaxy. Once again we found out that our Earth is far less important than we had believed, lying in an anonymous galaxy, not even an especially big one.

Over the next few years, different groups revised down the value of the Hubble constant—and increased the age of the Universe—time and again. What Hubble had thought were single stars in galaxies, for example, were actually clusters of stars. Again, Hubble had underestimated their luminosity and hence their distance.[21] In 1956 a study by Humason, Mayall, and Sandage suggested a value of 180 km/s/Mpc. Two years later, Sandage reported a value 75 km/s/Mpc. Over the years he was to lower the value still more. In collaboration with the Swiss astronomer Gustav Tamann, he produced a long series of papers suggesting that the Hubble constant was around 50 to 55 km/s/Mpc. Meanwhile, Sidney van den Bergh and Gerard de Vaucouleurs were giving values of just about double—around 100 km/s/Mpc—and ne'er the twain shall meet.

As John Huchra at the Harvard-Smithsonian Center for Astrophysics in Cambridge, Massachusetts, wryly pointed out, "The middle ground was littered with the bruised and battered remains of young astronomers attempting to resolve the dispute between the two sides." The scope of the problem can be seen in figure 5.4. In the left panel is a plot prepared by Huchra of all the published values of the Hubble constant. They reduce sharply and rapidly over the years until they become stuck between 50 and 100. In the right panel are values of the Hubble constant since 1970. There are well over 100. Values by different groups are identified: stars for Sandage and Tammann; pentagonal stars for Sidney van den Bergh and Gerard de Vaucouleurs; large dots for astronomers who would later use the Hubble Space Telescope to estimate the Hubble constant; and small dots for everyone else. To put it mildly, there has been violent disagreement.

With a Hubble constant of 50 km/s/Mpc, the Hubble time of the Universe would be 10 times that originally calculated by Hubble—30 Gyr. If its value is 100 km/s/Mpc, the Hubble time reduces to 15 Gyr. And, as the Universe is indeed not empty and subject to the gravity of all its content, the true age of the Universe would be significantly less.

That such a fundamental number for cosmology was known with an accuracy of a factor of two was scandalous. Something had to be done.

Hubble Strikes Back

In 1985 a group of frustrated cosmologists met in Aspen in Colorado. They suspected that there was as much chance of a white Christmas in the Sahara as there was of the rival groups agreeing on a value of the Hubble constant. Because the cosmologists also expected that the Hubble Space Telescope would be launched imminently,[22] they wanted to use it to resolve the problem once and for all. They agreed to prepare a major proposal to use the telescope. Because of its importance and the fundamental nature of its expected results, the proposal became a priority project for the telescope, formally identified as an HST Key Project.

The idea was to use the unprecedented capability of the Hubble Space Telescope to observe and measure Cepheid variable stars in distant galaxies, observations not possible from underneath the Earth's atmosphere. The aim was to measure the Hubble constant to a precision of 10 percent. As John Huchra later commented, "Examination of the 'error trees' for almost all previous determinations of H_0 showed that the nearly factor of two range in derived values was not unexpected given the large number of contributing parameters. At each rung of the distance ladder subtle and sometimes not-so-subtle choices introduced both larger and larger discrepancies and errors." The group decided to take a step-by-step approach to the problem from the very first rungs of what astronomers call the distance ladder—the distance to the Hyades cluster and to nearby stars—and to work in stepwise fashion measuring everything carefully right out to distant galaxies.

Observations with the Hubble Space Telescope started in 1994. Five years, 18 galaxies, and nearly 30 published papers later, the HST Key Project announced in the summer of 1999 that it had met its goal. The investigators reported $H_0 = 71 \pm 6$ km/s/Mpc. The result is not uncontroversial; the group led by Sandage and Tamann continues to hold out for a smaller value of H_0 (more like 60 km/s/Mpc), and consequently an older Universe, but most astronomers have been happy to accept the result. What is interesting is that the values have converged quite significantly and that most groups are now getting values fairly close to 70 km/s/Mpc. Even Sandage and Tammann now support a value significantly larger than the value of 50 km/s/Mpc they had defended for 20 years and one that is now closer to the Hubble result.

So, Just How Old Is the Universe?

If we take the HST Key Project result—71 km/s/Mpc—at face value, the Hubble time for the Universe is 13.8 ± 1.2 Gyr. This figure does not take into account that the matter in the Universe slows down the expansion—another highly controversial subject we will discuss elsewhere. Early in 2003, though, NASA announced the results from a new satellite. The Wilkinson Microwave Anisotropy Probe (WMAP), launched on June 30, 2001, observed the cosmic background radiation from when the Universe was just 380,000 years old. WMAP suggests the Universe is 13.7 ± 0.2 Gyr old, very close to the Hubble time for the Universe (see figure 5.5).[23] If true, this has important implications, given that the first stars must have formed extremely rapidly after the Big Bang, appearing within about 200,000 years of the formation of the Universe. It also puts very strong limits on the amount of ordinary matter in the Universe.

Perhaps new results will appear and overthrow these latest findings, but it seems that there is a nice agreement between the ages of the oldest stars, the Hubble time, and the results from the WMAP satellite. Maybe we are finally getting an accurate idea of how old the Universe is. Given the track record of such studies, however, perhaps we should not bet on it just yet.

SUGGESTIONS FOR FURTHER READING

Because our knowledge continues to change at a frightening pace, most published articles and books on the age of the Universe are of historical interest only. In this situation, the Internet has a huge advantage over other published media for giving up-to-date-information.

More Advanced Reading

E. Hubble, "A Relation between Distance and Radial Velocity among Extra-Galactic Nebulae," *Proceedings of the National Academy of Sciences* 15, no. 3 (March 15, 1929): 168–73.

> *Hubble's original article in which he established that the more distant the galaxy, the greater its velocity of recession, set down the Hubble Law, and demonstrated the expansion of the Universe. This article is available on the Internet at http://antwrp.gsfc.nasa.gov/diamond_jubilee/1996/hub_1929.html.*

A. Sandage, "Edwin Hubble, 1889–1953," *Journal of the Royal Astronomical Society of Canada* 83, no. 6 (1989).

> *To mark the centenary of Hubble's birth, the great modern astronomer, Allan Sandage, who at the start of his career collaborated with Hubble's assistant Milton Humason, wrote this biography detailing Hubble's work and achievements, which was published as a monograph. This article is available in the Internet at http://antwrp.gsfc.nasa.gov/diamond_jubilee/1996/sandage_hubble.html.*

On the Internet

The Time 100
www.time.com/time/time100/

> *The* Time *100 lists the most influential people of the twentieth century. Included alongside Albert Einstein and Jonas Salk (the inventor of the vaccine for polio that has almost eliminated this horrible disease from the world) is Edwin Hubble, the man who will be remembered for discovering the expansion of the Universe and for defining Hubble's Law that states how far a galaxy or quasar is from its red shift. Hubble's* Time *biography can be found at www.time.com/time/time100/scientist/profile/hubble.html.*

Hubble's Law
www.upei.ca/~physics/p221/pro99/hubble/hubble.htm

> *A site that contains a brief introduction to the Hubble constant and its importance and a series of commented links to other Web pages—including those that are generally considered to be the most important—that explain different aspects of the Hubble constant and its measurement. In particular, it notes the level of the different pages and what sort of public will be able to understand them in each case.*

The Hubble Constant
http://cfa-www.harvard.edu/~huchra/hubble/

> *John Huchra's site about the HST Key Project to measure the Hubble constant discusses the results that have been obtained. It contains fascinating historical background as well the history of the project and a series of links to other HST Key Project Web sites and to individual project results.*

CHAPTER 6

Is Anybody There?

Are we alone in the Universe? Are there other intelligent beings somewhere out there in space? Those are the questions I am asked most often when I give lectures and broadcasts. By turning on the television in almost any city on the planet, one can enter strange worlds where humans make daily contact with exotic aliens. Television series such as *Star Trek* remain astonishingly popular even 35 years after the short-lived show was officially killed off. So popular is the theme of humans gallivanting around the Galaxy, outwitting aliens, that it has led to a seemingly unending series of spin-offs like the later *Star Trek* movies and TV sequels, or spoofs, such as the film *Galaxy Quest* or radio series *The Hitchhiker's Guide to the Galaxy*, that range from the genuinely entertaining to the truly awful.

I have always been a voracious reader, and novels about contact with alien civilizations are often on my reading list. In many of them, the contact happens when the aliens decide to pay us a visit and take over our planet. I have yet to read a story in which the human race fails to beat off the invaders. Can we really expect to wage war with the inhabitants of the third planet of Arcturus with a guarantee that we will have the same weapons, or at least weapons similar enough to make it a good contest? Or are we more likely to find ourselves armed with metaphorical peashooters against their tanks and hydrogen bombs? Any war with aliens would be ridiculously one-sided—and, unlike our hero, Captain Kirk, we may not always be on the winning side! I have enjoyed the books in Harry Turtledove's World War series as much as anyone; their basic premise is that, shortly after the German invasion of Russia, a fleet of highly advanced lizard-like beings arrives from the star Tau Ceti and invades the Earth in an attempt to incorporate it into their empire. The aliens, of course, make it easier for the human race by using the same weapons that were already being used in the war, albeit better and more advanced versions of them, rather than hitting us with the double delta ray disintegrator or some other weapon so advanced that we could have no possible defense against it. But despite the spirited resistance offered by the human race in these and other similar novels, are we ever likely to best any invading aliens with no sense of property rights that might one day appear on Earth, when we are armed with only bullheadedness and Yankee ingenuity? Any civilization with hostile intent that is advanced enough to cross interstellar distances is likely to roll over our planet's defenses without breaking sweat, as do Tom Cruise's Martians in *The War of the Worlds*.[1]

Not all aliens in novels and science fiction stories are hostile. Some authors have populated our Galaxy with just a few peaceful civilizations and suggested that intelligence is rare in the Universe. A few have gone even further. In the Foundation Trilogy, Isaac Asimov painted a picture of a galactic empire formed purely and exclusively by humans who have swarmed over the Galaxy without ever encountering any kind of alien intelligence. Curiously, as we will see, the theory that the human race might be the only intelligent civilization in the Galaxy was proposed seriously in the 1980s. Can we then even be sure that we are not alone in the Galaxy? Are there any aliens out there to meet?

Some people will not only answer this question in the affirmative but also claim we are being visited regularly and that certain select humans have made contact and have even ridden in alien spaceships. These people are often termed the "Flying Sorcerers" and are a study in themselves.

Vulcans and Bug-Eyed Monsters

Many films and television shows depend on contact with intelligent aliens (*Star Trek, Star Wars, Babylon 5, Starship Troopers*, etc.). How likely is it, though, that such aliens with strikingly similar technological levels and (usually) remarkably human anatomy really exist? Could there be aliens with a chemistry different from our own, based, for example, on gold, or that can breathe ammonia, or live happily in a vacuum? Are we sure that our anthropomorphic attitude is correct? We have a big problem here. To date, the number of planets on which we know life has developed is just one: Earth. We cannot compare life on Earth with life anywhere else. Imagine growing up in Siberia and, with no knowledge other than what you see around you, having to work out what it is like to live in Paris. How would you do? Probably not too well—even though Siberians and Parisians are descended from the same hominids who roamed the plains of East Africa millions of years ago.

There are a few things that we can be confident about. Unless our ideas of biochemistry are erroneous, life must be based on complex molecules. Why does this rule out really exotic chemistry?

Making a Different Alien

One of the basic (bad) science fiction plots involves exotic beings based on some element such as gold. Why can we rule out such beings? The reason is that only carbon seems to be the base for life, although an alternative is possible. As life requires complex molecules—a molecule of DNA combines hundreds of thousands or millions of atoms—the element that is the base of life must be a good builder of molecules.

Few elements have a capacity for building molecules that are more than a few atoms in combination. The best builder element is carbon. There are various reasons for this, but perhaps the most important is that carbon can form as many as four bonds with different atoms, with the priceless ability to form single

$$C - C,$$

double,

$$C = C,$$

or even triple bonds

$$C \equiv C$$

with other carbon atoms, or with other elements. This gives it a capability to form a huge variety of molecules, combining with many other elements. Proteins, one of the basic molecules that make life possible, require a complex structure of carbon, oxygen, nitrogen, and hydrogen that is only possible because the carbon atoms act as a link between other elements that are *not* good builders of large molecules.

In the periodic table, an element that lies immediately above or below another will have similar properties. Scientists call molecules formed using these similar elements "chemical analogues," which is just a way of saying that they are the same molecule with the exception of the substitution of one atom for a similar one. The element below carbon in the periodic table, which has similar properties, is silicon. There is bad news, though, for people who speculate about silicon-based life forms. Silicon can form large molecules, but only by the simple repetitions of atoms. For example, animals breathe out carbon dioxide, one of the simplest of organic molecules. The silicon analogue is silicon dioxide, the molecule that forms sand or glass. Why is carbon dioxide a gas and silicon dioxide a solid that melts at high temperature? The reason is that the molecules of silicon dioxide form huge chains with each other. Unfortunately, these chains are repetitions of silicon and oxygen atoms, rather than the complex combinations of various elements that carbon forms. Silicon dioxide is poisonous when breathed in as dust, causing silicosis, although silicon is a vital element in the human body helping to form bones.

Whatever we may read in science fiction stories, the existence of silicon-based life forms is unlikely, though not quite impossible.[2] What about other carbon analogues? The list of elements similar to carbon and silicon is surprising. In order of increasing mass, they are germanium, tin, and lead. Germanium is a well-known semiconductor, often used in electronics. Tin and lead are widely used because they are relatively inert. Thus, we use tin as a coating on cans to preserve food, and the Romans used lead for pipes (however, unlike tin, lead is poisonous and so one line of speculation is that the Roman empire decayed due, at least in part, to the progressive lead poisoning of its population from the water pipes). These elements do not build large molecules at all and are, unlike carbon, unwilling to combine with anything. So dream of lead-based organisms if you wish, but they are not going to exist.

It is now easy to see why talk of exotic life forms is so unrealistic. If even the second-best builder element after carbon is a poor substitute and other carbon analogues are even worse, it is not sensible to talk of other elements that do not share chemical properties with carbon as even remote possibilities; nature and the laws of science do not allow it.

Jim, It Is Logical That I Should Not Exist

Even if it is with a small "if, and, or but," it seems probable that any life forms we will encounter in the future will be carbon-based. That does not mean that they will necessarily be similar to us. But, even if all life is carbon-based, could a half-human, half-Vulcan like Mr. Spock ever be born?

It seems ridiculously unlikely that a completely alien planet will have hit on DNA as the basis for life. It is such an extraordinarily complex molecule, and there must be millions of possible variants on it. Unless other planets have life forms astonishingly similar to our own, there would be no possibility at all of cross-breeding. Probably we could not even eat, or rather get nourishment from, the same types of food (to be able to do that, the protein molecules would have to be similar to those found in life on Earth — possibly we could *eat* alien foods without being poisoned, but would slowly starve to death, however much we ate, because it would simply not be usable by our body). But to give birth to Mr. Spock, even more is required: 96 percent of the genetic code of humans is the same as that of apes, but the 4 percent of difference makes it impossible for humans and apes to mate successfully. Even on Earth with its enormous similarities between species, cases where one species can mate with another are rare. Is it really reasonable to believe that we could mate with an extraterrestrial with a completely different body chemistry, not to mention anatomy, and which may not even limit itself to just two sexes?

To quote Carl Sagan, speaking before his untimely death, "We have as much chance of being able to crossbreed with an extraterrestrial as I have of mating with a petunia."

As Mr. Spock would say, "It is not logical that a half-Vulcan could one day exist."

Cosmological Enigmas

Drake and the Dolphins

In a universe of 100,000 million galaxies, each composed of 100,000 million stars, can we be alone? Are the starways crisscrossed with messages between one planet and another? Do the spaceways heave with ships crossing the distances between the stars? Could there be hyperintelligent godlike civilizations out there that built the Universe in the way that Carl Sagan suggested in his novel *Contact*? To answer such questions, we examine how life developed on Earth and extrapolate this to life on other planets. We must examine the famous Drake equation and look at how modern research has affected the numbers that come out of it.

In November 1961 a historic meeting was held at Green Bank Observatory in West Virginia. The Space Science Panel of the National Academy of Sciences convened 11 distinguished scientists. Apart from radio and optical astronomers, there was a physicist and a biochemist. One member, a researcher into the chemistry of the origin of life, Melvin Calvin, heard during the meeting that he had won the Nobel Prize for his work. John Lilley, an expert in the language of dolphins, rounded off the team. Because of him, the group christened themselves the Order of the Dolphin and later received dolphin badges to mark their participation. The meeting was to discuss extraterrestrial life and how we might communicate with it—hence the presence of an expert in a nonhuman language.

Drake formulated the equation that would center their efforts:

$$N = R^* {}^* f_p {}^* n_e {}^* f_l {}^* f_c \cdot L.$$

This means that N, the number of civilizations that may be able communicate with us, is the product of R^*, the rate at which Sun-like stars were being formed at the time that our Sun formed; f_p, the fraction of stars with planets; n_e, the number of planets per solar system suitable for life; f_l, the fraction of suitable planets on which intelligent life develops; f_c, the fraction of intelligence civilizations that desire to communicate; and L, the lifetime of each communicating civilization.

Although many of the factors were uncertain, their conclusion was that the number of communicating civilizations in our Galaxy would be about the same as the lifetime of each one. In other words, if an average civilization lasts just 100 years from the time that it starts to be able to communicate, there may be just

100 civilizations in our Galaxy and thus the nearest would be so far away (10,000 light years or more) that we would have great difficulty in finding it. In contrast, if civilizations last 1 million years each on average, the nearest might be very close to us—perhaps just 100 light years away.

At the time the Order of the Dolphin first met, several of the numbers were totally speculative. For example, not a single extrasolar planet was known; thus one could only guess how many stars might have planets. Even today it is not possible to answer this question reliably. Similarly, the number of planets per solar system suitable for life was based on little more than a guess. Here, recent research suggests that their estimate was highly optimistic to suggest that as many as five planets per solar system might have conditions to develop life.[3]

How Life Started

The oldest fossils on the planet are found, logically enough, in the oldest rocks, although that mere fact actually gives us some important information. Scientists talk of 1,000 million years (an American billion) as an aeon. Our solar system formed approximately 4.7 aeons ago. Perhaps 4.5 aeons ago the Earth was destroyed in a giant impact and reformed along with the Moon.[4] The oldest rocks are 3.8 aeons old and contain fossils of single-celled algae. Thus at some point in a space of just 700 million years, the Earth resolidified and cooled, the oceans formed, and life developed. This suggests that in some sense the formation of the first life forms was in some way easy, but how did they come about?

In the early 1950s, Melvin Calvin at the University of California at Berkeley and Harold Urey and Stanley Miller at the University of Chicago experimented with jars of a mixture of gases that reproduced the primitive Earth's atmosphere. They bombarded the mixture with radiation or with sparks to simulate lightning. In doing so, complex organic molecules were formed. The different groups assumed that the primitive Earth's atmosphere would be similar to that of Jupiter, with large amounts of hydrogen, ammonia, methane, and water vapor. The results were extremely encouraging. After just a week of "cooking," as much as 15 percent of the methane had been converted into organic compounds that included two amino acids: alanine and glycine. Other experiments produced even more complex molecules. This led to a belief that life probably started on Earth in a chemical soup in the oceans. Some estimates suggested that the amount of organic

Cosmological Enigmas

chemicals produced by reactions in the atmosphere would be so large that as much as 1 percent of the content of the oceans would have been organic material.

Some scientists had their doubts. One important problem is that there is no evidence for this organic soup at all. If the oceans did consist of this organic soup, one would expect evidence of it in ancient rocks, but there is none. Later, the whole scenario was cast into doubt. Scientists realized the Earth's iron core makes the presence of a large amount of methane unlikely in the primitive atmosphere, because this methane would combine with the iron. This makes it likely that the Earth's atmosphere was made up largely of other gases and that building complex organic molecules would be much more difficult. Could the appearance of life on Earth have had outside assistance? We already know that we are in one important sense products of processes that take place in deep space[5]—many of the atoms in our bodies were formed in other stars—but could it be that the organic compounds that make up our bodies were also formed in space? Back in the 1950s, Spanish biochemist Joan Oro made a bold suggestion that only recently has started to be taken seriously: many of the organic chemicals that formed the basis of life on Earth came from comets.

Some meteorites contain complex organic compounds, and it was later discovered that such compounds are also common in comets and in interstellar space. Dozens of organic molecules have been detected in space, including relatively complex compounds such as formamide (CH_3NO), which contains carbon, oxygen, nitrogen, and hydrogen, and the current largest known interstellar molecule, the 13-atom cyanodecapentyne ($HC_{11}N$), a long chain of carbon atoms with a hydrogen and a nitrogen atom at either end.

$$H - C \equiv C - C \equiv C - C \equiv C - C \equiv C - C \equiv C - C \equiv N$$

The presence of so many organic compounds of moderate complexity hints that there are probably smaller amounts of very much more complex compounds. It could be that the impact of comets on our Earth has had much more fundamental effects than could have been imagined even 20 years ago.

Over the past few years, we have come to realize that we may owe the existence of the oceans on our planet to the water that has fallen to Earth in comets. Comets, though, bring far more than just water. In the 1970s Fred Hoyle and Chandra Wickramasinghe suggested that life might have developed in space and that the constant drizzle of meteoric dust from space could seed the Earth with new viruses. They also suggested that certain features seen in the spectra of molecular clouds in space could be consistent with the existence of simple bacteria in those clouds. Few scientists accept this suggestion, but there is increasing evidence that cometary nuclei may be rich in relatively complex organic molecules. These molecules form naturally with the bombardment of frozen gases adsorbed into dust grains on the surface by cosmic radiation. The same process explains the formation of organic molecules in dust and gas clouds in space: a rich soup of gases binds to the dust grains and is transformed by radiation into increasingly complex molecules.

There is good reason to believe that even if life on Earth did not come from comets, its initial appearance was given a great deal of assistance by the fall of organic material from space each time a comet collided with our planet. In other words, we are even more "children of the stars" than is generally realized. Maybe life started on Earth and was extinguished several times over a period of tens or hundreds of millions of years until it finally took hold.

What Does a Planet Need?

In our solar system there are eight planets as recognized by the International Astronomical Union, but of them only Earth is known to be an abode for life. Lowly organisms may exist on Mars.[6] More speculative are suggestions that there may be organisms in the oceans of Europa and the satellite of Jupiter; more recently, two scientists at the University of Texas have proposed that microbes may exist in the clouds of Venus.[7] There has even been a suggestion that the origin of the methane in the atmosphere of Titan may be bacteria in the interior of the satellite.[8] Even if we do find that simple organisms exist elsewhere in the solar system, it is evident that only the Earth is a suitable abode for more advanced life, but why?

When the Order of the Dolphin was meeting, it was thought that any rocky planet within a fairly broad band of distances around a medium-sized star like the Sun would be a suitable abode for life. The width of the band was decided by the range of distances from a star where liquid water could exist: in other words, between 0 and 100°C. In our solar system, Venus and Mars are just inside the limits of what was termed the Sun's biosphere. The panel suggested that as many as five planets in a single solar system might be suitable for life. As I suggested previously,[9]

Cosmological Enigmas

this is extremely optimistic. The Earth is almost exactly centered in Sun's theoretical biosphere, but were the Earth's orbit a little more eccentric or a little closer to the Sun, it is almost certain that no life would have been possible on our planet. It is mere chance that the temperature of our Earth is "just right." It is pure chance that the Earth is not so warm that a far more violent greenhouse effect takes place, or so cold that the oceans freeze over, robbing our planet of enough greenhouse effect to keep it warm. In other words, probably the most critical item to make a planet suitable for life is that its distance from its star is such that it has just the right amount of greenhouse effect in its atmosphere to permit the existence of temperate oceans on its surface.

When we look at the planets of other stars, we see that cases like our Earth may be far less common that we had previously thought.

Planets of Other Stars

Until the mid-1990s there was no direct proof of the existence of planets around any star other than the Sun. The evidence for extrasolar planets—planets of other stars—was tenuous. However, as of November 25, 2006, there are now 197 confirmed extrasolar planets orbiting 169 stars. More are being added every month, although what makes something a planet is open to debate, as there is no official definition. Some experts count any object too small to permit the simplest nuclear fusion reaction (deuterium burning) as a planet. This includes objects to be as large as 13 times the mass of Jupiter or 1.2 percent of the mass of the Sun. Objects several times the mass of Jupiter are usually called super-Jupiters, and not all astronomers accept them as genuine planets; rather they are regarded by some as being failed stars.

The first confirmed extrasolar planet was announced in 1995 by Michael Mayor and Didier Queloz. It orbits the star 51 Pegasi. The planet was detected with the Elodie spectrograph at the Observatoire de Haute Provence, measuring the tiny movement of the star caused by the pull of gravity of the planet as it moves in its orbit. The big surprise was that the planet is very close to the star but similar in size to Jupiter. This led to it being dubbed a "hot Jupiter." With an orbital period of only 4.23 days, different estimates put the temperature of the surface at between 950°C and 1150°C, hot enough to melt aluminum, copper, or gold.

To date we know of some 80 planets similar in size to Jupiter that orbit their parent star at a distance similar to, or less than, that of Mercury from the Sun and that are classed as "hot Jupiters" (see figure 6.1). Almost all the planets detected to date have caused consternation because they are so unlike our own solar system. Where there is no hot Jupiter, the planet is almost always in an eccentric orbit. In the case of the star HD80606, its planet is several times the mass of Jupiter and has an orbit not unlike that of Halley's comet. A large planet in so eccentric an orbit would destabilize the orbit of any other planet in that solar system, making it unsuitable for life.

The most common techniques used to detect extrasolar planets work best at detecting large planets close to a star, so it is not surprising that the majority of planets discovered so far are unusual when compared to our own solar system. To detect a planet, one has to observe it for at least as long as it takes to orbit once around its parent star. To detect Jupiter, then, an astronomer in a distant solar system would have to observe our Sun for at least 12 years. This is difficult, but in July 2003 there was some excitement in the astronomical community when astronomers at John Moores University in Liverpool announced that they had detected a new solar system that is the most similar to our own (see figure 6.2). The new planet is twice the mass of Jupiter and has a circular orbit with a period of 6 years around HD70642, a seventh magnitude star similar to the Sun, in the southern constellation of Puppis. Such "normal" planetary systems must be common too, even if more difficult to detect, but we have no idea how common. Many experts argue that all stars that are not part of binary or multiple systems—about half of the stars in the Galaxy—will have planets. Even if only half of these are normal planetary systems, their number will exceed 10,000 million. A recent study cast grave doubt on this number. No fewer than 754 nearby stars have been carefully examined for the presence of planets; this is a large enough number to allow astronomers to reach some important conclusions. Of these stars, 61 had detected planets and 693 have no detectable planet.

When astronomers looked at the composition of the stars with and without planets, they made a surprising discovery: the stars that had a small metalicity—that is, that had only a small amount of elements heavier than helium and are thus very old[10]—have few detected planets. In contrast, according to a study announced in summer 2003 by Debra Fischer of the University of California, Berkeley, and

Jeff Valenti of the Space Telescope Science Institute in Baltimore, Maryland, a staggering 20 percent of young stars in the sample have detectable planets. Younger stars with a much larger metal content are thus five times as likely to have had planets discovered around them. This means that we can begin to suspect that the fraction of stars with planetary systems in the Galaxy may be much less than 50 percent because, if only the younger stars are likely to have planets, we must exclude many stars as being potentially planet bearing.

How Many Civilizations?

Even taking the most pessimistic estimates used by the Order of the Dolphin, as many as 100 intelligent civilizations might be expected in the Galaxy. This in itself creates an important problem that was the subject of a heated debate during the 1980s and 1990s. The argument goes something like this: if there are a significant number of intelligent civilizations, then even if they only expand through the Galaxy at velocities consistent with rocket flight, they should have filled the entire Galaxy by now. However, there is no credible evidence of extraterrestrials either on Earth or in the solar system; if other civilizations exist, they should have visited us by now. They do not appear to have done so, suggesting they do not exist.

This argument is difficult to counter. Some scientists talk about "God's quarantine regulations," suggesting that interstellar distances are so large that the vast distance between inhabited planets stops civilizations contacting each other. Others say perhaps they have come but have not left evidence on Earth: if one wants to find space-faring civilizations, one leaves the calling card where only beings that have conquered space travel will find it. It is thus suggested that evidence of extraterrestrial visits is most likely to be found on the Moon (as in *2001: A Space Odyssey*) or in the asteroid belt. As the human race has barely started to explore either, even a large monument may take many years to locate. Another widely touted possibility is that perhaps there is a galactic "Prime Directive" and that all contact is prohibited until a fledgling civilization has shown itself to be ready for contact. Perhaps they are out there, know that we exist, but have no intention of revealing themselves quite yet. This is in turn countered by the argument that, if there were many civilizations in the Galaxy, surely there would be maverick civilizations that would ignore these rules of isolation.

Clearly, we cannot make any reliable guesses about the probability of finding other civilizations in the Galaxy on the basis of the lack of evidence that we have so far. There are too many unknown quantities involved. It is generally tacitly assumed that the development of intelligence is almost inevitable if life appears on a suitable planet. But is it? Earth managed without it for some 4,000 million years and recent studies show that the human race came close to extinction, perhaps dropping to just a few thousand survivors at one point. Some writers have even suggested that the critical factor in developing intelligent life on Earth has been the high rate of mutations caused by the wide distribution of radioactive elements in the Earth's crust. If true, the Earth may be unusual in developing advanced plant and animal life forms in a relatively short time. We have also seen that Drake and his colleagues may have vastly overestimated the number of planets suitable for life in the Galaxy and very probably the number of stars with planets. Until we start to get information on the number of Earth-like planets around nearby stars, attempting to estimate how many civilizations are out there may be a pointless exercise. Were we to discover that instead of there being a "one planet per star" minimum suitable for life, the actual number is one planet per 100 Sun-like stars, then the lower limit given by the Drake equation for the number of civilizations in our Galaxy drops to one: us. We might genuinely be alone in the Galaxy.

ET, Phone Home!

The ultimate aim of the Green Bank conference was to discuss the possibility of communicating with extraterrestrial civilizations. At the time, Frank Drake was involved in Project Ozma, named for the Wizard of Oz. Its purpose was to listen for possible intelligent radio signals from the two Sun-like stars closest to the Earth—Epsilon Eridanii and Tau Ceti—using the radio telescope at Green Bank. This project was the longest of long shots with the technology of 1960, but such attempts to search for intelligent signals from space were for many years known as "CETI" (communication with extraterrestrial intelligence) in honor of Tau Ceti.[11] Over the following 15 years, no less than eight other major projects to look for intelligent extraterrestrial signals were launched in the United States and the Soviet Union. Some focused, like Project Ozma, on individual stars. Others, such as a search by Frank Drake and Carl Sagan with the Arecibo telescope, concentrated on observing whole galaxies in the hope that one might have a super-transmitter capable of sending signals over millions of light years. Still other searches looked for signals from all over the sky. There have been many different projects aimed at detecting extraterrestrial signals. The best known is probably the Seti@Home project. Seti@Home uses the combined computing power of nearly 4 million computers around the world, most of them PCs at home, to search for possible intelligent signals in data recorded by the giant Arecibo radio telescope in Puerto Rico (see figure 6.3).

Of all the different projects, one has produced an almost legendary event. On August 15, 1977, Dr. Jerry Ehman, a scientist at the Ohio State University Radio Observatory, noticed an astonishingly strong signal from a point in the sky in the constellation of Sagittarius that was being observed with the "Big Ear" telescope. Amazed, he wrote the word "WOW!" alongside it on the computer printout (see figure 6.4). There are many misconceptions about this signal. Although it has no known explanation, it was most definitely not a message as such; it was a single, short-lived blast of noise from a rather localized point on the sky, with no content of information. This point in the sky has been reobserved many times, and the signal has never repeated. No known human agency produced the signal, but that does not mean that it was necessarily extraterrestrial in origin. It probably has a perfectly innocent explanation, but we just do not know what it is!

In recent years there have been many wide-ranging SETI projects. To date there have been more than 60 SETI programs, but most of them have been small-scale and involved little time actually observing the sky and searching for signals. Among the most ambitious have been Project META and META II. META began in 1985 using a 26-meter radio disk belonging to the Harvard-Smithsonian Institution to map the whole sky at wavelengths around 10.5 and 21 centimeters. Although 37 suspect signals have been observed, none have repeated when the position was reexamined. META-II uses two 30-meter radio telescopes of the Instituto Argentino de Radioastronomia near Buenos Aires (Argentina) for a similar project covering the southern sky. More recently the META project was replaced by BETA, a greatly upgraded search with the same telescope. Another major project has been SERENDIP, which started in the late 1970s. Run by Berkeley, successive versions—SERENDIP I, II, III, and IV—have been mounted on different telescopes, listening in while they take data for other projects. Since April 1992, SERENDIP III has been mounted at Arecibo and, as astronomers take data, it has analyzed the noise from space in search of potential signals.

To date, no intelligent extraterrestrial signals have been found. Perhaps they never will, but the implications of the receipt of an intelligent signal from space are so vast that it is worth the effort.

. . . but Does ET Already Know Us?

Although some effort has been made to listen to possible signals from space, almost no effort has been made to send signals. Perhaps everyone is listening and nobody bothers to transmit? Whether we like it or not, we reveal our presence. Since the 1920s, radio and television signals have been broadcast that have leaked into space. Over the past 50 years, these transmissions have become very much stronger and could be detected from other stars with sensitive-enough equipment. Right now, for example, Neville Chamberlain's speech announcing the start of World War II has reached, thanks to the BBC, 64 light years into space and passed hundreds of stars. News of Jimmy Carter's election reached Vega in 2002 and is heading onward and outward. On Tau Ceti, anyone listening to broadcasts from Earth would have heard of the first Gulf War at the start of 2003. What impression would an extraterrestrial have of life on Earth from watching our TV programs?

There is good news and bad news. Reruns of *I Love Lucy* and *Scooby Doo* are not the most powerful signals being sent into space. Far more powerful transmissions are coming from the chain of great radar stations across the Northern Hemisphere. Although they are not broadcasting messages, they would undoubtedly be recognized as artificial signals. Astonishingly, some of these radars would appear brighter than the Sun when observed from another star. Intelligent extraterrestrials may even recognize them for what they are, which is not necessarily a good thing. Perhaps these signals have not yet reached another civilization. But if they have, they would reveal a lot about the human race, not all of it good. Maybe ET does know all about us and has decided to leave well enough alone!

Take Me to Your Leader!

If alien civilizations do exist in the Galaxy, could they have visited us, or even be visiting us right now?

One of the hazards of being an astronomer is the constant questioning about

Cosmological Enigmas

"aliens." Frequently I receive calls from members of the public (these calls always seem to come to me) after they have seen something odd and want to know what it was. In most cases I can give a fairly satisfactory explanation by application of common sense and a reasonable knowledge of the sky. More complicated are the small and dedicated minority who see flying saucers nearly every day. Such people not only know that they have witnessed alien spacecraft but, in some cases, profess to have spoken to the aliens on board.

Some cases are clearly fraudulent: one flying saucer classic shows what professes to be a "mother ship with five smaller craft." Any astronomer will rapidly identify the photograph as being an overexposed image of the planet Saturn taken through a telescope large enough to show its main moons. The photo is often presented, though, as having been taken by a flying saucer enthusiast (actually it was taken by a famous astronomer with a well-known telescope in a major observatory in Texas). A few years ago there was a huge flying saucer watch organized at Los Llanos de Ucanca at an altitude of 2,200 meters (7,500 feet) at the base of Mount Teide on Tenerife. Many thousands of people attended, including a fair number of astronomers, but the only person to report sighting something was the organizer of the watch himself, whose photograph of the spaceship, most unfortunately, did not come out!

It is important to remember the difference between a flying saucer and an unidentified flying object (UFO). Use of the term flying saucer implies that an object is something extraterrestrial—to date, I have not registered a single flying saucer, nor do I expect ever to see one. UFOs are another matter. Anyone who watches the sky attentively will see occasional UFOs. When I am asked if I have seen one, I always reply that I have on several occasions: a UFO is simply any object in the sky that the observer cannot identify. Thus, many phenomena that the average member of the public may see and not be able to identify are classic UFOs. In the 1970s even President Jimmy Carter saw a UFO and requested that NASA carry out an investigation into the phenomenon;[12] what the president had seen was just the planet Venus, but he did not know that at the time.

Almost invariably when someone calls to report a bright light in the sky, it is Venus (around dawn or dusk) or Jupiter (late at night). If Mars is in opposition in summer, as it was in 2001 and 2003, I can expect a spate of calls about a bright red light in the sky—Mars—which will be even brighter than Jupiter for a few weeks. On other occasions a bright light in the sky has been a bright meteor burning up in the atmosphere. One caller recently reported seeing an unusual, glowing cloud like an aircraft contrail in the night sky: this appeared to be the smoky trail that a very large and bright fireball must have left in the sky after it had burned up. Only a few astronomers will ever have seen this phenomenon and, even though it is completely natural, it is exceedingly difficult to recognize what it is without specialist knowledge.

Other UFOs are more mundane. Local legend on the island of Tenerife states that there is a flying saucer base at a place called Montaña Roja (Red Mountain), an extinct volcano on the coast alongside Tenerife's southern airport. The quantity of sightings of strange lights around the mountain makes Tenerife one of the UFO hotspots of Spain and of Europe. The mountain is alongside a major international airport with its radar, which should detect alien spacecraft as effectively as charter flights full of British and German tourists, and which has an inviting runway. Flying saucers are neither detected in flight, though, nor do they land in the space provided alongside the terminal building. What the flying saucer enthusiasts do not mention is that the isolated beach by the Red Mountain is reputed to be a favorite site for landing contraband at night, particularly tobacco. In other words, the lights around the mountain are of greater interest to the local police and customs authorities than to an astronomer.

Many UFO sightings are most certainly not natural phenomena. Military operations, by their natural need for secrecy, cause many UFO reports. Some classic Spanish UFO cases have turned out to be tests of submarine-launched missiles out in the Atlantic. Experimental aircraft and even weather balloons can cause convincing reports. In a field where rumor and misinformation are rife, sometimes the military prefers an incident to be treated as a UFO rather than say what was really seen. A case in point is the reported incursions by Cuban MiG fighters into the Mississippi valley in the 1960s and 1970s. No U.S. military commander would ever wish to announce that a UFO sighting was actually evidence of a hostile aircraft flying unmolested over the United States—at least, not if he wants to keep his job. So it is logical to deny that U.S. Air Force aircraft were in the region at the time and let people draw their own conclusions.

I have seen what appeared to be flying saucers on a number of occasions. A couple of times, my heart started racing until I realized that my eyes were playing tricks on me. At the time I lived close to an airport with an approximately east-west runway. In the afternoon the small turboprop aircraft that mainly use the airport often fly into the Sun, producing brilliant reflections. The observer's eyes tend to ignore the thin wings and register what they are seeing as a brilliant round or saucer-shaped object. Unless you watch carefully you can be convinced that you are seeing a flying saucer. Most times one suspects that imagination and desire to see and believe do the rest. I suspect that many "flying saucers" are due to similar optical illusions, which are difficult to recognize in the heat of the moment.

As for those people who assure us that flying saucers have landed and that they have spoken to the occupants, ask yourself how likely it is that if an alien civilization did pick our Sun to visit out from among the 100,000 million stars in the Galaxy, they would speak perfect English! Is it not more likely that their anatomy would be so different from ours—can we really expect them to have human vocal cords?—that we would probably be incapable of even recognizing their speech as such and they would be incapable of reproducing the sounds that we make. If you believe the more extreme flying sorcerers, Earth must be the JFK airport of the Galaxy, even though our planet is remote, circling an undistinguished star in a distant part of one of the outlying arms of the Milky Way. If other civilizations are capable of crossing interstellar distances, there must be many far more interesting places to go.

Gods, Monkeys, and Men

The number of civilizations in our Galaxy is sufficiently low that it is extremely unlikely that we will ever meet a civilization close to our own technological level. In any encounter with extraterrestrials, we will either be hundreds or more likely thousands of years ahead or behind our contactees. All such encounters will thus, at best, be like a contact between our twenty-first-century civilization and that of ancient China, or the Egypt of the Pharaohs. One possibility then is that our hypothetical Captain Kirk would feel as if he has moved back in time to an era equivalent to ancient history, or to the dawn of civilization. In this case he would be like unto a god, and the locals will not understand him at all. The other is that he would be like the poor pyramid constructor suddenly and rudely transported into modern New York City, unable to communicate with the locals and driven mad by what he sees. Arthur C. Clarke once said famously that when we venture into space, "We will meet monkeys or angels, but not men."

Whomever we meet, they will be so far ahead of us there is a real danger we will not be able to understand them, or so far behind that we will not want to. Some experts have pointed out that there may be grave dangers in contact. This does not mean that extraterrestrials may try to take over our planet: interstellar conquest would almost certainly be ruinously expensive and unlikely to be worth-while—any potential alien invaders would be so far ahead of us in technology that we would have no possibility of stopping them. Our experience on Earth, though, has been that when an advanced civilization meets a less-developed one, the less-developed civilization almost invariably comes off worse and, in many cases, has disappeared. Perhaps an attempt to find ET is too risky for our own survival.

SUGGESTIONS FOR FURTHER READING

Popular Books

Fred Hoyle and Chandra Wickramasinghe, *Life Cloud* (London: Sphere Books, 1979).

In this book, Hoyle and Wickramasinghe propose their theory relating comets and molecular clouds with life on Earth. There are some highly controversial comments on the possible extraterrestrial origin of cold and influenza viruses. The book is worth reading, but many of its conclusions are not generally accepted by scientists.

Science Fiction

Contact with extraterrestrial civilizations is a common theme in many novels and stories. Some are humorous, some attempt to investigate the possible consequences for our planet and the human race, and many have unfriendly aliens who try and take over the Earth. Here are a few suggestions for reading that contain some good, realistic "what if" science.

A. C. Crispin, *Starbridge* (New York: Ace Books, 1989).

The author of the well-known television series V has written a series of books about human contact with a galactic federation in the distant future when interstellar travel has become widespread and easy. The basis of the book is that there are other civilizations in the Galaxy, but there are very few of them. An intelligent and thoughtful book about a first contact and its possible consequences.

Larry Niven and Jerry Pournelle, *Footfall* (London: Sphere Books, 1985).

This is one of the many novels that treats an unfriendly first contact. As in many cases, the alien invaders are defeated by human ingenuity and determination. This book is worth reading because every effort is made to make the science believable and as accurate as possible, although, as usual, the alien technology is of a level close enough to ours that the battle is unrealistically even. This successful collaboration has produced various classic science fiction novels.

Harry Turtledove, *World War in the Balance* (London: Hodder & Stoughton, 1994).

Another case of alien invasion, this one occurs at the height of the Second World War. One of the interesting aspects of this book, the first in a series of six novels to date, is that exploration of "what if the human race is in some way different in its development." This is mixed with a highly authentic historical setting exploring what might have happened if aliens had arrived at a critical moment in human history.

More Advanced Reading

NASA, *The Search for Extraterrestrial Intelligence,* edited by Philip Morrison, John Billingham, and John Wolfe (New York: Dover, 1979).

A NASA publication based on a series of workshops on extraterrestrial communication held in the mid-1970s. It included contributions by various distinguished scientists. The book provides details of possible search strategies, panel discussions, and technical requirements for searches. Although there is some technical discussion and a few equations, it is a good review of the theory behind CETI that, to a large degree, is still valid even now, almost 30 years on.

On the Internet

Los Alamos National Laboratory's Chemistry Division's Periodic Table of the Elements: http://pearl1.lanl.gov/periodic/default.htm

This simple and highly practical Web site allows you to click on any of the 112 known natural and manmade elements and learn their properties and even their cost. The details on carbon explain what makes this the fundamental element in all life forms. Compare what is said about silicon with the properties of carbon.

A Home from Home
www.pparc.ac.uk/Nw/Md/Press/HomeFromHome.asp

This press release announced the discovery of a new solar system similar to our own around the star HD70642. The images that accompany the press release and that include a spectacular artist's impression of the new planet and an animation of the journey to the new solar system are at the following address: http://www.pparc.ac.uk/Nw/Artcl/images_to_accompany_press_releas.asp.

Astrobiology in New Scientist
www.newscientist.com/hottopics/astrobiology/astrobiology.jsp

This fascinating compilation of many articles on astrobiology, the science of life on other worlds, is reprinted from New Scientist *magazine, an authoritative British science magazine read by scientists and the public alike.*

Seti@Home
http://setiathome.ssl.berkeley.edu/index.html

These pages give information about the project Seti@Home, the opportunity to download the program and look for signals at home on your own computer, and a breakdown of results and analysis. It is essential reading.

The WOW signal
www.bigear.org/default.htm

This page is dedicated to the Big Ear Radio Observatory in Ohio, which was finally retired in 1997 and dismounted in 1998. This observatory features in the Guinness Book of Records *as having the longest running SETI program in the world. The pages include photographs about the observatory and detailed information about the WOW event.*

http://demoprints.eprints.org/archive/00000453/01/The_Seti_WOW_Signal.pdf

This fairly technical document discusses the WOW event in detail and possible explanations of it.

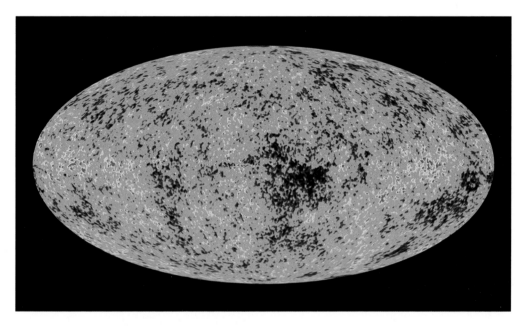

Figure 5.5. The map of the cosmic microwave background produced by the Wilkinson Microwave Anisotropy Probe (WMAP). In the image, red indicates areas of the microwave background that are a little warmer, and blue, areas that are colder. The light that is being detected in the image was emitted just 379,000 years after the Big Bang. This map has allowed cosmologists to measure the age of the Universe with greater precision than ever before and also to discover that the first stars in the Universe formed more quickly than had previously been thought.

Figure 6.1. An artist's impression of the planet of 51 Pegasi burning close to its parent star.

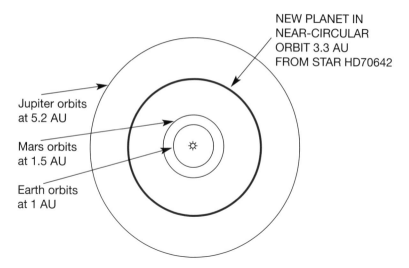

NEW PLANET IN
NEAR-CIRCULAR
ORBIT 3.3 AU
FROM STAR HD70642

Jupiter orbits
at 5.2 AU

Mars orbits
at 1.5 AU

Earth orbits
at 1 AU

Figure 6.2. A comparison of the planet of HD 70642 with the orbits of Mars and Jupiter. The new planet is 90 light years from the Sun. Billed as the planetary system most similar to ours, it is one of the few planetary systems that we know that is "normal" in our experience. AU = Astronomical Unit (E/arth-Sun distance).

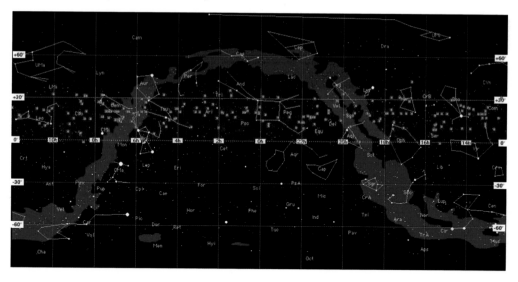

Figure 6.3. A sky map showing the best signal candidates in the Seti@Home search, shown as red and yellow squares. A total of 216 stars and possible extraterrestrial signals were reobserved with the Arecibo telescope between March 18 and 24, 2003, to check them for possible signals.

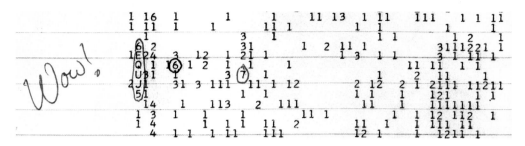

Figure 6.4. An image of the paper chart showing the WOW event and Dr. Jerry Ehman's reaction to it. This is a map of a small area of sky observed by the Ohio State University "Big Ear" radio telescope. Numbers and letters show the strength of the signal from each point in the sky as it scanned in front of the telescope. No number or a 1 indicates no signal or a very weak signal, and an increasingly large number represents a stronger signal. In this case the strength of the signal reached 31. The recording plots numbers up to 9 and then continues with A (10), B (11), C (12), etc.

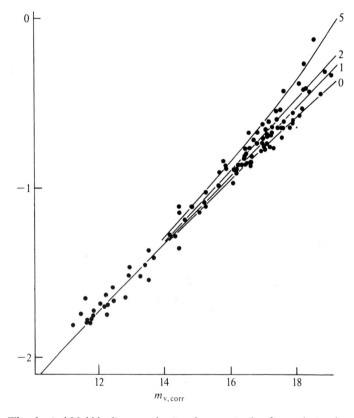

Figure 7.1. The classical Hubble diagram plotting the magnitude of 103 galaxies that are each the brightest in a cluster of galaxies against the logarithm of the red shift (in other words, a red shift of 0.01 is plotted as −2, one of 0.1 as −1, and one of 1.0 as 0). Four theoretical curves are shown for different values of q. Unfortunately, the dispersion of the data is such that it is impossible to distinguish which of the lines the data follow. This problem has haunted all such efforts to measure q.

Figure 7.2. A Hubble Space Telescope mosaic of part of the Coma Cluster of galaxies. Note the brilliant giant elliptical galaxy in the upper left quadrant of the image that is so much brighter than any other galaxy in the field of view.

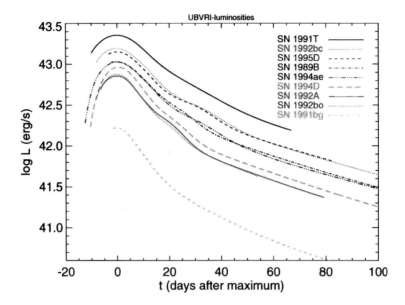

Figure 7.3. The light curves of nine supernovae of type Ia showing how the shape of the light curve changes according to the luminosity of the supernova, with the most luminous fading considerably more slowly that the faintest.

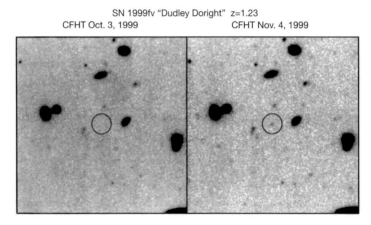

Figure 7.4. The supernova Sn 1999fv, discovered in images taken with the 3.5-meter Canada-France-Hawai'i Telescope at Mauna Kea in Hawai'i. The before (*left*) and after (*right*) images show the difference between October 3, 1999, when the supernova was invisible, and November 3, 1999, when it was discovered. This is typical of the faintness of some of the supernovae discovered by the High-Z Supernova Search Project. The red shift of the supernova ($z = 1.23$) corresponds to about 10,000 million light years distance.

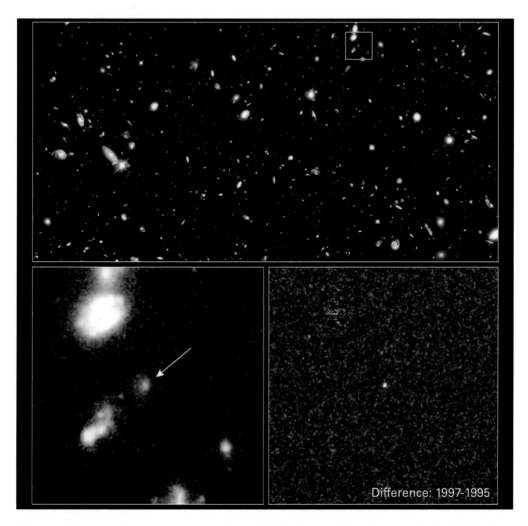

Figure 7.5. The most distant supernova yet detected. This object was found in images by the Hubble Space Telescope in 1997. At a red shift of z = 1.7, it is at a distance of 11,300 million light years. In the different panels we see the original image from the Hubble Deep Field in 1995, the image taken in 1997 where the position of the supernova is marked (the orange blob is the supernova's galaxy, not the supernova itself), and the final panel shows the 1995 image subtracted from the 1997 image, showing the small, faint point of light that was the supernova.

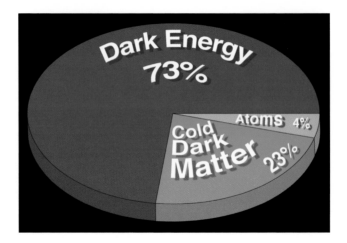

Figure 7.6. How the Universe is built. Almost three-quarters of the Universe is made of the mysterious "dark energy," the antigravity force that causes the expansion of the Universe to accelerate, with almost all the rest "cold dark matter"—black holes or exotic particles that we cannot detect directly.

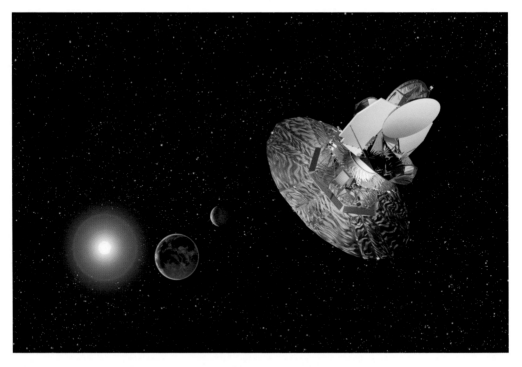

Figure 7.7. The Wilkinson Microwave Anisotropy Probe (WMAP), which has confirmed that the expansion of the Universe is accelerating and thus that it will continue forever.

Figure 8.1. Lunar surface images taken by the Apollo XI astronauts. The image above, by Neil Armstrong, shows Buzz Aldrin setting up the lunar seismometer. The image on the next page depicts the view from Aldrin's window after the moonwalk, showing the flag, footprints, and various craters and boulders. Note how totally black the sky is. No stars are visible because the exposures given are only of the order of 1/100 second to avoid overexposing on the brightly lit lunar surface and are thus too short to record stars.

Figure 8.2. A panoramic image of the Milky Way from our perspective in space drawn in the 1950s by a team of artists under the supervision of Knut Lundmark at Lund Observatory in Sweden. Seven thousand stars have been painted in by hand, as have the star clouds. Note how the form of the Milky Way is broken by numerous dark patches that are due to dust clouds in the plane of the Galaxy.

The APM Galaxy Survey
Maddox et al

Figure 8.3. The APM (Automated Plate Measuring) map of the distribution of galaxies in the sky. Note that there are hints of a honeycomb structure with more galaxies in some directions and fewer in others. The APM Galaxy Survey is a computer-generated sky survey of more than 2 million galaxies, covering about one-tenth of the whole sky, in the South Galactic Cap, made by Steve Maddox, Will Sutherland, George Efstathiou, and Jon Loveday, with follow-up by Gavin Dalton.

Figure 9.1. The Hubble Space Telescope image of the field of NGC 4319 and Markarian 205 in approximately true color. Note that many faint and distant galaxies can be seen in the image, at least two of which can be seen clearly shining through the disk of the galaxy. The galaxy that hosts Markarian 205 is clearly visible and partially superimposes on the spiral galaxy. The bright starlike object just to the lower left of Markarian 205 is another galaxy at the same red shift as the quasar.

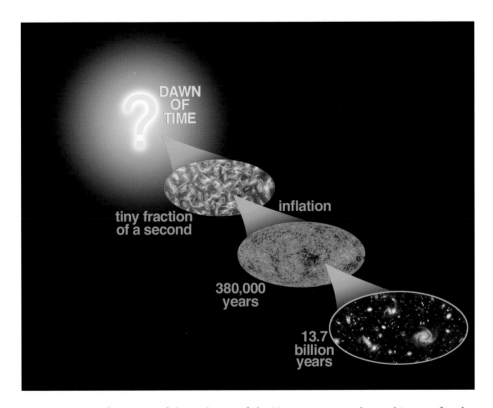

Figure 9.2. A brief overview of the evolution of the Universe as we understand it now after the latest cosmological discoveries.

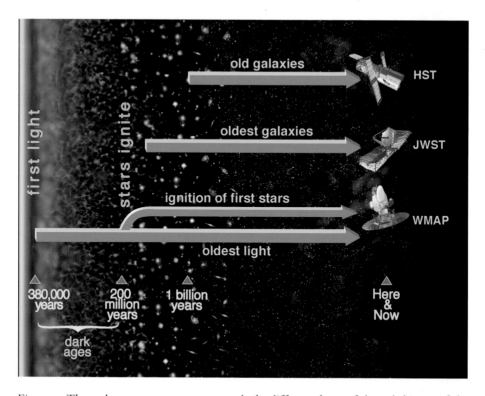

Figure 9.3. The tools astronomers can use to study the different phases of the early history of the Universe. Satellites like the Wilkinson Microwave Anisotropy Probe are the only tool that allows us to study the so-called dark ages before the formation of the first stars. Before the moment when the Universe became transparent, there is an impenetrable wall that blocks our view of earlier times. The Hubble Space Telescope can study extremely old galaxies but does not have enough reach to see the first stars and galaxies of all after the dark ages ended; astronomers hope that its much-larger replacement, the James Webb Space Telescope (previously known as the Next Generation Space Telescope), will be able to see right back to the formation of the first galaxies soon after the first stars formed. The Herschel Space Observatory will be able to study star formation in some of the oldest galaxies.

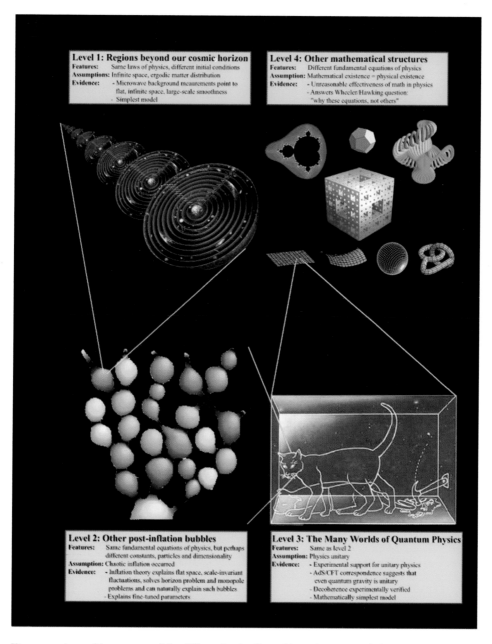

Level 1: Regions beyond our cosmic horizon
Features: Same laws of physics, different initial conditions
Assumptions: Infinite space, ergodic matter distribution
Evidence: - Microwave background measurements point to flat, infinite space, large-scale smoothness
- Simplest model

Level 4: Other mathematical structures
Features: Different fundamental equations of physics
Assumption: Mathematical existence = physical existence
Evidence: - Unreasonable effectiveness of math in physics
- Answers Wheeler/Hawking question: "why these equations, not others"

Level 2: Other post-inflation bubbles
Features: Same fundamental equations of physics, but perhaps different constants, particles and dimensionality
Assumption: Chaotic inflation occurred
Evidence: - Inflation theory explains flat space, scale-invariant fluctuations, solves horizon problem and monopole problems and can naturally explain such bubbles
- Explains fine-tuned parameters

Level 3: The Many Worlds of Quantum Physics
Features: Same as level 2
Assumption: Physics unitary
Evidence: - Experimental support for unitary physics
- AdS/CFT correspondence suggests that even quantum gravity is unitary
- Decoherence experimentally verified
- Mathematically simplest model

Figure 10.1. A possible structure of the different levels of possible multiverses, with their distinguishing characteristics.

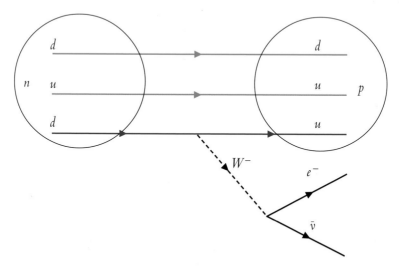

Figure 10.2. A Feynman diagram of beta decay, which is the decay caused by the weak nuclear force. In the diagram we see how a neutron (*left*, labeled n) with d (or down) quarks and a u (or up) quark is changed into a proton (*right*, labeled p) when the weak force intervenes and a d quark turns into a u quark by emitting a W particle (known to physicists as a W boson), which decays into an electron and an antineutrino. In a universe in which the weak nuclear force is stronger, beta decay would happen more easily, and radioactive elements that decay by beta decay would be more radioactive and shorter-lasting.

CHAPTER 7

How Will the Universe End?

The past few years have been fascinating ones for cosmology because we are getting closer to a definitive answer to one of the
greatest and most intractable problems in this field. The explanation that is emerging is so unexpected that shock waves are still re-
bounding through the scientific community as it attempts to fit the new
information into our ideas of the Universe.

Two fundamental questions are how was the Universe born and how
will it die. For the first, we think that we now have the complete and
definitive answer; for the second, the issue is far trickier. Answering the
fundamental questions requires that we first address a whole series of
more specific issues: Did the Universe always exist? Will it always exist?
Will the Universe continue to expand forever until all its stars fade out
and die? Or will it stop expanding and collapse in on itself? If it collapses, will a new Universe form from the ashes of the old one?

Speculation about the end of the universe inevitably has something
of an emotional content; I have always found the idea of a Universe that
expands forever until the last red dwarf star has faded away to be a most
depressing thought, even though such a fate would be hundreds of

(American) billions of years in the future. As we will see, in the past the rival steady state theory found a neat escape clause in this sad scenario by supposing that matter in the form of hydrogen atoms is spontaneously created, very slowly, to fill the void created by the expansion. Almost nobody now supports this model of the Universe, but it was a way of making the Universe everlasting, by allowing it to keep making new stars to replace the old ones.

Until recently, the eventual fate of the Universe was an extremely controversial topic; now it has been downgraded to simply controversial. Estimates of the rate at which the expansion of the Universe is slowing down (the "deceleration parameter") require some 10 times more matter to be present than is detectable in the visible Universe. This gives rise to theories of dark matter filling the space between the galaxies. Different measures of the deceleration parameter gave widely differing values, but a consensus seemed to be forming that the true value was close to the critical value at which the amount of mass closes the Universe at infinite time. Theories of the Big Bang seemed to suggest that this was a natural value. Recently though, new observations have thrown things into complete confusion and, once again, have proved the old adage "the Universe is not just stranger than we imagine, but stranger than we *can* imagine." Not just that, but a supposition described some 75 years ago by Einstein himself as "my greatest ever blunder" has returned to haunt us long after his death.

Setting the Scene

After the Second World War, astronomers were starting to think seriously about the consequences of the expansion of the Universe after the theories of stellar evolution began to give a solid idea of the lifetime of stars. Stars are born, do die, and have finite lifetimes, so adding the expansion of the Universe into the picture made astronomers wonder: what happens next?

The Big Bang theory was first proposed, although not under that name, in 1927 by the Belgian priest Georges Lemaître, who had independently derived the Friedmann-Lemaître-Robertson-Walker equations and suggested, based on the red shifts of galaxies observed by Slipher, that the Universe began with the explosion of what he termed a "primeval atom." Hubble's work on the expansion of the Universe made the Lemaître model the logical choice. This model was later ac-

Cosmological Enigmas

cepted and strongly advocated by George Gamow, who is the scientist most associated with it.

Not everyone accepted this model of a Universe with a definite beginning and, by implication, a definite end. In 1948 three eminent scientists—Fred Hoyle, Thomas Gold, and Hermann Bondi—proposed a second and radically different idea to explain the expanding Universe. They proposed that the Universe has always existed and always will. Although it is expanding, new matter, in the form of hydrogen atoms appears spontaneously to fill the void between the separating galaxies. This leads to the pleasing symmetry that the Universe has always and will always have the same appearance, because, as stars and galaxies die, new ones will replace them.

During the 1950s and 1960s the steady state theory had a strong and loyal following among astronomers. Although it probably never became the theory supported by the majority of astronomers, a substantial minority accepted it. When the BBC broadcast its documentary *Violent Universe* in 1969—a major production with a running time longer than two hours that aimed to give the viewer an eyewitness account of the new astronomy of the late 1960s—some tens of scientists were asked to give their views as to which of the two theories they supported in an informal poll. Although the steady state theory of the Universe was definitely losing support in the scientific community, a substantial minority of those asked for their opinion still favored it at that time.

Part of the theory's appeal was undoubtedly that it offered a simple and optimistic view of the Universe. The theory stated that the Universe has always existed, always will, and would always remain basically identical; it avoided some difficult questions, in particular how, when, and why the Universe began. It also painted a picture of constant regeneration, although at such a slow rate that it could never be detected. In a galaxy the size of the Milky Way, just a few hundred new hydrogen atoms would have to be formed each year to make up for the expansion. These few hundred atoms would allow the Universe to have eternal life.

No explanation was offered for this creation of mass and how it might occur, which was one of the weaknesses of the theory—it could only say what its supporters thought was happening, not explain the physics of how it happened. In the end, this weakness killed it; the steady state theory could not adapt and made rigid predictions about the Universe, being unable to explain in a convincing way the new discoveries that astronomers made in the 1960s. Some efforts have been made to revive the theory and a small minority of astronomers, such as Halton Arp, point to objects in the sky that they suggest may be generating matter that is being expelled into intergalactic space, but very few astronomers supported them and their number has reduced still further in the past 15 years.

The overwhelming majority of astronomers now regard the whole subject of the steady state theory with some distaste. In this sense, science is unforgiving rather in the same way that politics is: once you are seen to have failed, unless something wholly unexpected and spectacular happens, making a comeback is almost impossible.

So, if everlasting Universe must be rejected, just how and when will the Universe end?

Looking for q

Just like in a James Bond movie, in cosmology "M" and "q" call the shots and provide the tools that the cosmologists use to solve the case when they are spying on the Universe's secrets.[1] So, what are these mysterious agents and how did they come about?

All mass has gravity, and the more mass there is, the stronger the force of gravity. And gravity pulls things back—this is the famous (and totally erroneous) phrase, "what goes up must come down." Take a ball and throw it up in the air, and it will rise a certain distance, slowing all the time, and eventually fall back. Throw the ball harder, and it will take longer to slow down and stop and thus will get higher in the air but, sure as eggs are eggs, it *will* come back down. Use a mortar to launch the ball, and it will go even higher, but it will still come back down. So why is it not true that "what comes up must come down"? Because if you throw the ball hard enough, it will go so fast that, even though the Earth's gravity will slow it, it will never stop and fall back. This was the principle of the Apollo Moon flights. They built up such a high speed that, even though the Earth's gravity tugged them back, they were still traveling at about 4,500 kilometers per hour when the Moon's gravity took over and made them accelerate once again. The Pioneer and Voyager spacecraft built up such a high velocity that even the Sun's gravity could not stop them, and they are headed out into interstellar space, still traveling at high speed. In other words, if you go fast enough to reach the escape velocity of a body like the Earth, you will escape from it and never fall back, even if at the end you are slowed to a crawl.[2]

What Does This Have to Do with q and the Fate of the Universe?

The Universe is evidently not empty (which is a pity, because it *does* complicate things for the cosmologists). Thus gravity will slow down its expansion. If you can measure how much the expansion of the Universe is slowing down, then you can weigh it. Back in the 1950s, this parameter in the equations was christened "q" by the cosmologists who wanted to measure the expansion of the Universe. So everything boiled down to finding something that one could measure at increasing distances and see how much it varied and thus measure q and solve the problem in an instant.

What is q and what do the values mean? Well, if the Universe were empty and its expansion not slowing down, its value would be zero and the relationship between red shift and distance would be a perfect straight line. The Universe would expand forever and never stop. However, the line bends because the mass of all the stars and galaxies slow the expansion at a rate that depends on how massive the Universe is and thus how big q is.

As shown in table 7.1, if q = 0.5, the Universe is just massive enough that after an infinite amount of time the expansion will stop.[3] For this to happen, we only need the Universe to average about six atoms of hydrogen per cubic centimeter over its entire volume, which is an astonishingly small density.

If q is greater than 0.5 by however little—in other words, if the average density of the Universe is more than six hydrogen atoms per cubic centimeter—the fate of the Universe is sealed: gravity will win the tug of war with the headlong rush of recession of the galaxies, and all the galaxies will eventually rush back to their point of origin. Instead of a "Big Bang," there will be what astronomers call the "Big Crunch." The Universe as we know it will be destroyed, potentially to explode again in a new Big Bang and to form a new Universe.

Table 7.1.   Important Values of q and Their Meaning

q	Result
−1	Steady state model correct: An *everlasting* Universe
0	Empty Universe with no mass whose expansion will never slow: An *open* Universe
0.5	Universe just massive enough to stop expanding at infinite time: A *critical* Universe
1	Massive universe that will stop expanding and will contract again afterward: A *closed* Universe

But How to Measure q?

In the 1950s astronomy was advancing at a great pace. Two things seemed to offer good possibilities of measuring q. The first was the huge advances being made by using the new 200-inch Hale reflector at Mount Palomar. With it, astronomers had a tool that allowed them to see further out into the Universe than ever before and see fainter and more distant objects. The second was the rapid advances being made in the new science of radio astronomy.

As we will see, initially astronomers were very optimistic about radio astronomy and the possibility that it offered to solve quickly the mystery of the fate of the Universe. However, as has so often happened when we have investigated deep questions of the Universe, things turned out to be much less simple that they appeared to be initially.

It is time to look at the way that astronomers first tried to solve this problem, the difficulties that they found, and how they tried to overcome them until, quite literally, they saw the light.

Radio Eyes to the Universe

The history of radio astronomy can be traced back to the end of the nineteenth century and the experiments that Nikola Tesla carried out with "electric emissions," but it was not recognized as an astronomical tool until after the Second World War. Karl Jansky had shown that radio waves came from the Sun and the Galactic Center. The only person who had followed up the discovery, and that some 10 years later, was effectively an amateur astronomer, Grote Reber,[4] which shows how much impact Jansky's work had made in reality. All that was to change, though.

In the mid-1930s, although the top level of the British government seemed remarkably unconcerned, not everyone viewed developments in Europe so calmly. There was an increasing concern in some sectors of society about German rearmament and Hitler's evident expansionist aims. This was manifested in two important ways. First, the British government started a register of scientists and their experience. This meant that when war eventually did break out, some 8,000 scientists were known and identified. As a result they could be put to use at any moment in war research projects,[5] and each scientist could be assigned to the project that most suited his experience and abilities. Second, there was considerable worry about the possibility of a Nazi "death ray" that could be directed at aircraft and either destroy them or, at least, kill the pilot. Various scientists were put to work on the problem. It soon became obvious that a death ray as such would be impossible. It would require such high levels of energy that it would simply not be feasible. However, scientists noticed a second possibility: that an airplane would reflect a significant amount of energy back to the transmitter and so reveal its presence even when not visible to the naked eye. Tests revealed that, even with the simplest and most primitive equipment, a bomber could be detected at a range of some kilometers and its approach tracked.

In the end, Sir Robert Watson-Watt, who developed the first system to carry out radio detection and ranging of aircraft, ran the project. Scientists could now detect an aircraft at a range of tens of kilometers, well before it could reach the transmitter. It calculated the aircraft's range—how far away it was—and its altitude with some precision, especially when the operators had some practice in interpreting the oscilloscope traces.[6]

After the war there were hundreds of decommissioned scientists who had worked with radar and huge amounts of war-surplus electronics that the government was willing to sell. In late 1945 one of the scientists, Bernard—later Sir Bernard—Lovell, filled a trailer with equipment that he had bought for five pounds (about ten dollars at the current exchange rate, but then more like twenty-five dollars) and headed off across the fields to some land that the University of Manchester owned on the plains of Cheshire, with the idea of using it to study radio emissions from the sky. The trailer got stuck in the mud some way before reaching its objective, and the scientists had to accept that they would get their precious equipment to move no further and would have to set up shop where they were. The equipment worked to such good effect that, eventually, the spot where the trailer became bogged down became Jodrell Bank Radio Observatory. Of such trifles are history made.

Radio astronomers discovered increasing numbers of sources of radio waves in the sky, what they called "radio stars," but even as late as the early 1950s few of them could be identified with known objects because their position was only poorly known—radio telescopes give such a blurred view of the sky that it was extremely hard to locate objects accurately enough for an optical telescope like the Hale Telescope to be able to find them. This is simply a consequence of the way that radio telescopes work: the resolution of a telescope depends on its size (more is better) and the wavelength that it uses (shorter is better). So, even though the 76-meter telescope at Jodrell Bank was 15 times larger than the Mount Palomar 200-inch telescope, the radio waves it uses have a wavelength about 100,000 times greater than visible light, so the images from Mount Palomar are about 10,000 times sharper (or, in other words, have 10,000 times more resolution).

A few identifications were obvious—the Sun, the Galactic Center, Jupiter (first detected by accident in 1955), the Andromeda Galaxy, the Crab Nebula—but most were a mystery, and the nature of radio sources such as Cygnus A, Centaurus A, and Virgo A was unknown. Bright stars seemed not to emit radio waves, and astronomers realized that, if the Sun were moved out to the distance of the stars, its emissions would be too weak to detect.

The breakthrough came in 1951 when a young researcher at Cambridge's Cavendish Laboratory, Francis Graham Smith, applied a new technique, one that combined two radio telescopes to give a much more accurate position. He measured positions for Cygnus A, Puppis A, and Cassiopeia A that were accurate enough to make it worth turning the Hale Telescope to that point of the sky. Smith sent his measures to Walter Baade and Robert Minkowski, who turned the Palomar telescope to them in September 1951. Cassiopeia A was a faint nebula in our Galaxy—later identified with Tycho's supernova of 1572. Cygnus A was much stranger. Baade and Minkowski found an odd-looking galaxy with a red shift of 16,830 kilometers per second at that position, which appeared to be two galaxies in collision (we know now that it is a nearly spherical galaxy whose center is almost completely hidden by a thick band of opaque dust that makes it appear, from a distance, to be split in two).

As radio observations improved in quality, it was found that the emissions from Cygnus A came from two distinct and symmetrical blobs, with the galaxy found by Baade and Minkowski exactly in the middle of the two. Many other similar radio sources were found, with two large lobes of radio emission on either side like the giant ears of some cartoon caricature. Oddly, though, the galaxy seen in the visible images could not be detected with the radio telescopes, and the giant ears of radio emission could not be seen with optical telescopes.[7]

Knowing that Cygnus A and later Centaurus A were very distant galaxies, astronomers wanted to use them to measure the Universe. If all were like Cygnus A, then the further away the galaxy, the smaller and fainter it would be. It was calculated that, depending on how the expansion of the Universe was slowing down at great distances, their sizes would indicate the rate of slowing down. If the steady state theory were true, mathematical analysis would show that the sources would reach a minimum size of eight seconds of arc at great distances. Other models made different predictions, so the observations would tell astronomers which of the models was correct.

Because of the problem of the inevitably blurred images, no single radio telescope could make these measurements on its own. The solution was to combine two telescopes separated by a certain distance to form a single instrument equivalent in size to the separation between them. This technique is known as interferometry.

Various groups set out to find out how small the different sources were. As they were relatively brilliant, it was not necessary to use a big telescope. At Jodrell Bank they decided, for example, to combine the main telescope—known as the Mark I—with a small, portable antenna. By observing at a wavelength of 21 centimeters with two telescopes separated by just 6.6 kilometers, it would be possible to measure the diameter of a radio galaxy eight arcseconds across. The Jodrell Bank astronomers set out with enthusiasm and started to measure different radio sources to see if they could be resolved—that is, detected to be larger than the resolution of their instrument—with their two telescopes. Each time they found that some of the sources resisted being resolved, they began the laborious process of moving the portable telescope to a new and more distant location and trying again.

After a time frustration began to set in. The portable telescope was far more than 6.6 kilometers away, but a number of the radio sources stubbornly refused to cooperate. The scientists turned to other solutions, such as using another radio telescope that was available more than 60 kilometers away, but to no avail. Some of the radio sources were much smaller than expected,[8] and because the results showed radio sources in general are of very different sizes, such generalization did not work. Sadly, this method of calculating q failed; not all sources were exactly like Cygnus A.

It was time to fall back on Plan B.[9]

Further and Fainter (or, If Only Life Were That Simple)?

The alternative method for calculating q depended on measuring the brightness and red shift of objects and seeing how they varied. Measure 10 galaxies at different distances and you will find that the most distant is the faintest. In other words, a plot of the magnitude against the red shift will show a nice line. As the red shift gets larger and larger, and thus the objects are at greater distances, that line should start to bend depending on how large the value of q is. Astronomers estimated that by a red shift of 0.8 the bending *should* be detectable, and at larger red shifts it would be easy to measure (see figure 7.1).

It rapidly became clear that this method was running into a fundamental problem. In this case, it was M. A plot of the brightness of galaxies against their red shift would work perfectly, provided that all the galaxies have exactly the same luminosity. Astronomers usually define the luminosity of objects by their absolute magnitude—the magnitude that they would have in theory if we could see them from a distance of 10 parsecs—and call this value M. If M is constant, or almost constant, for all the objects in your plot or Hubble diagram, you get a nice line or curve. If M is not the same for each object, your Hubble diagram gets more and more confused according to how much difference there is in M from object to object.

It was not a great surprise to know that galaxies are not all exactly the same. Just looking around our Milky Way, we find galaxies of all sizes and luminosi-

Cosmological Enigmas

ties. There are three spiral galaxies in our own Local Group—the Milky Way, the Triangulum Spiral, and the Andromeda Galaxy—and some 30 to 35 other galaxies, some irregular in shape like the Magellanic Clouds, others tiny dwarf elliptical galaxies. Of these, the Andromeda Galaxy is by some way the largest and most massive—300 billion (3×10^{11}) times the mass of the Sun. It is followed by the Milky Way, 200 billion times the mass of the Sun: the Large Magellanic Cloud, 25 billion times the mass of the Sun; and the Triangulum Spiral, 8 billion times the mass of the Sun. In contrast, the dwarf elliptical galaxy Leo II has just a million times the Sun's mass.

When we use galaxies of any kind, straight out of the grab bag, although it is obvious that they get fainter as the red shift increases and the galaxies get further away, there is so much variation between galaxies that we cannot say much more than that. We need to use just one kind of galaxy that we can rely on to be constant.

We see the same when we look out into space. Nearby we find two huge clusters of galaxies: the Virgo cluster and the Coma cluster (see figure 7.2). Virgo has more than 2,000 large galaxies, which rather puts our poor Local Group to shame. When we look at any large cluster of galaxies, though, we find a giant elliptical galaxy in the center. Giant ellipticals—their name comes from their size and their shape—are the largest galaxies in the Universe and may be as much as 100 times more massive and luminous than the Milky Way.

This gave astronomers the possibility of using these galaxies as a "standard candle"—that is, to assume with some reasonable expectation of success that all giant elliptical galaxies are pretty close to the same luminosity, as if every giant elliptical galaxy is marked like a light bulb, "guaranteed 100 Watts."

The results, while better than for just any old galaxy, were disappointing. The example shown in figure 7.1 already comes from just such a process of carefully selecting galaxies: in this case, the brightest galaxies in more than 100 large clusters of galaxies. Although by a red shift of 0.5 the curves start to separate, the brightness of the galaxies separates even more. It is impossible to distinguish a tendency in the graph.

If one takes the average of all the galaxies, the value of q comes out as $q = 1.6 \pm 0.4$,[10] which would close the Universe very comfortably—it would have more than three times the mass necessary to stop its expansion. However, nobody really believed this result because there are two obvious problems. First, not all the galaxies have the same luminosity—they are no better than fair as "standard candles." Second, there is no guarantee that a giant elliptical galaxy today is exactly the same luminosity as a giant elliptical galaxy several thousand million years ago; we know that galaxies are systems that change and evolve with time, and it is likely that over thousands and millions of years their brightness varies. In other words, even if our giant elliptical galaxy has a big sign on it that says "guaranteed 100 Watts," there will probably be some small print warning "when new only—this value can change with time."

Cosmologists have tried many ways of solving this problem and getting information about the evolution of the Universe. One that was attempted for many years was to count the number of galaxies at different red shifts and see how the number varied in the past. Again, the idea was that if the expansion of the Universe was slowing at great distances, space should appear more squashed up and so fewer galaxies will fit in it. They had already tried adding up the masses of all the galaxies that were visible in the Universe and found that it was less than a tenth of the number that would be needed to stop the expansion—rather than six hydrogen atoms per cubic centimeter throughout the Universe, there appeared to be only about 0.5—which made it obvious that a lot of the mass in the Universe was hidden, the so-called dark matter. Dark matter is any kind of matter that does not make up stars and shine. This could be thin gas between the galaxies, or a large invisible cloud of cold gas surrounding normal galaxies in a kind of halo, or large numbers of black holes—anything that we could not see.

The measurements were difficult. Many effects had to be taken into account and corrected as much as possible. Everything suggested, whatever way the calculations were done, that the Universe was *open* and so its expansion would never stop.

Supernovae to Probe the Cosmos

For cosmologists, the past few years have provided many of the answers we have been seeking. They have been exciting times, but the answers have been unexpected and disconcerting.

The breakthrough came in the mid-1990s when astronomers realized there was a good standard candle that had never been recognized. There are at least four types of supernovae, known as Ia, Ib, Ic, and II. The last three all seem to be the explosion of massive stars, but the Ia supernovae are much more interesting. The currently favored theory is that they are caused by the implosion of white dwarf stars in a binary, or double-star, system.[11] Matter falls onto the white dwarf from the normal star and accumulates. However, a white dwarf star has a maximum possible mass called the Chandrasekhar limit, 1.4 times the mass of the Sun, above which it is no longer stable. When a white dwarf reaches that mass, it collapses into a neutron star. There is a consequent massive nuclear deflagration as all the material in the star suddenly combines in an orgy of nuclear reactions. There are good reasons for thinking that the Chandrasekhar limit is constant throughout the Universe, and so supernovae of type Ia are all of the same luminosity.

In 1993 Gustav Tammann suggested that this similarity between type Ia supernovae existed in reality. In the same year, Mark Phillips at the Cerro Tololo Interamerican Observatory in Chile noticed that even though it was not true that all were the same luminosity—unfortunately it turns out that the most luminous are about a factor of two brighter than the faintest—a fainter supernova fades much more rapidly than a more luminous one. It is therefore possible to work out exactly how bright the supernova is and compensate by observing carefully the changes in its brightness.

In other words, if you observe a type Ia supernova and measure its light curve carefully to see exactly how rapidly it fades from the maximum of the explosion, you can work out with great precision how luminous it is really. Figure 7.3 shows how this works with 13 supernovae observed from Cerro Tololo. In the top panel you can see how the shape of the light curve—the variations in the brightness of the supernova—changes between the most luminous and the least. In the bottom panel you see how, when this effect is taken into account and compensated in the light curve, all 13 supernovae superimpose almost perfectly. Type Ia supernovae do genuinely seem to be the perfect standard candles that astronomers had hoped for.[12]

Once you have your standard candle, you have to use it well. That means finding as many as possible as far away as possible.

Up to the early 1990s supernova hunting was a slow business. No more than a couple of dozen were found each year, most of them in nearby galaxies. In the mid-1990s a series of automated supernova search programs were set up. These started to find supernovae in ever greater quantities. In recent years the numbers of supernovae discovered have increased at an amazing rate. In 2004 a total of 249 supernovae were discovered, of which an impressive 120 were of type Ia.[13]

By 2006 the record score of 2004 had doubled to a barely credible 511 supernova discoveries.

Amateur astronomers with CCD cameras make many supernova discoveries. Among the supernovae discovered in the last few years, one finds many such discoveries. Recently, the British amateur Mark Armstrong, for example, has found 14 in 18 months to add to his many discoveries over the years. Another British amateur, Tom Bowles, has 28 in the same period of time. American amateur Tim Puckett has 21. However, the most successful search for supernovae in the world is an automated professional one: the Lick Observatory Supernova Search (LOSS) uses a 76-centimeter telescope atop Mount Hamilton (San Jose, California) to search for supernovae.[14] Since starting in 1997, the telescope has now discovered 605 supernovae (to late April 2007), with 95 in 2003 alone (its best year). A professional program like LOSS takes images of dozens of galaxies each night, which are then compared to an archival image of the same galaxy. The computer compares the two images and looks to see if there are any differences. It subtracts the archival image of the galaxy from the new one, and any starlike image that remains is a potential supernovae that can be confirmed with a new image. Using this technique, LOSS has found supernovae as faint as magnitude 20, although most are in the range from magnitude 16 to 18.

To get information about the future of the Universe, we need to find even fainter supernovae in very distant galaxies and compare their observed brightness to the expected brightness. Various teams, among them the High-Z Supernova Search Project and the Supernova Cosmology Project have been looking for extremely distant supernovae using large telescopes.

Many high red shift, and thus extremely distant, supernovae have now been discovered. The High-Z Supernova Search Team christens each supernova with a private name apart from the official International Astronomical Designation; for example, all the supernovae discovered by the team in 2001 were named after dinosaurs, while discoveries in 2000 were named after cartoon and film characters (Wonder Woman, The Incredible Hulk, Spiderman, Batman, etc.).[15] Figure 7.4 shows an example of one of the faint supernovae that the team has discovered. Sn 1999fv was 1 of 20 supernovae discovered with the 3.5-meter Canada-France-Hawai'i Telescope at Mauna Kea in Hawai'i in two nights on November 2 and 3, 1999, and rejoices in the nickname of "Dudley Doright." This supernova was magnitude 24.5 and was not even the faintest of the objects discovered with the tele-scope on those two nights.[16]

What was exceptional about Sn 1999fv was its distance. Its red shift of $z = 1.23$ puts it at a distance of about 10,000 million light years, one of the most distant supernovae ever detected.

The most spectacular search and the claim to having found the most distant supernova ever almost inevitably involve the Hubble Space Telescope. In December 1995 it observed a field of view in the Northern Hemisphere for 10 consecutive days reaching close to magnitude 30. In October 1998 another field was observed in the Southern Hemisphere.[17] Later, astronomers reobserved the same area to see if any supernovae could be detected in the images. Two candidate objects were detected in the later images, both of which were confirmed to be supernovae and designated Sn 1997ff and Sn 1997fg.[18] Of these, Sn 1997ff was magnitude 27 and found to be at the unprecedentedly high red shift of $z = 1.7$, putting it at 11,300 million light years distance (see figure 7.5).

This supernova confirmed a sensational result that had recently been announced from studying other distant supernovae; although the faintest and most distant supernova ever detected, it confirmed that the Universe is suffering from inflation.

Inflation and a Definitive(?) Answer

In 1998 astronomers from the High-Z Supernova Search Team and the Supernova Cosmology Project made a sensational announcement. By observing 16 supernovae with red shifts between 0.16 and 0.62, they found that the distant supernovae were 10 to 15 percent fainter than expected, showing they were more distant than could be explained by any standard model of the Universe.

Careful analysis of the observations showed that the data could only be fit by using a value of q smaller than the one that an empty Universe would give. In fact, the value that came out of the analysis was that $q = -1.0 \pm 0.4$. Given that the steady state theory had long been eliminated on other grounds, this could only mean that the Universe was more exotic and unusual than anyone had imagined.

Excluding one uncertain supernova, the observations also gave the age of the Universe as 13.6 Gyr, with a range of values from 12.8 to 14.6 Gyr. This result is extremely close to the one calculated in a completely different manner from later observations from space of the Wilkinson Microwave Anisotropy Probe (WMAP), which suggests very strongly that this is the true age of the Universe.

The distant supernovae were fainter than expected even from an empty Universe, but we know that the Universe is not empty. The best estimate that can be made is that its density is about 30 percent of that necessary to close the Universe (about two atoms of hydrogen per cubic centimeter). In other words, despite the fact that gravity should be pulling the supernovae back quite substantially and making them nearer and *brighter* than expected, they are actually *fainter* and more distant.

How Do We Resolve This Paradox?

The answer is that the Universe is full of a type of antigravity force called dark energy. What this does is to counteract the effects of gravity at great distances and to push the galaxies apart much faster than the expansion of the Universe would predict. This dark energy makes the expansion of the Universe accelerate rather than slow down.

The results from WMAP and high red shift supernovae have allowed cosmologists to calculate exactly what the composition of the Universe must be. The recipe for the Universe is 4 percent atoms; 23 percent cold dark matter (exotic material that cannot be detected directly—this matter may be black holes or some kind of exotic particles with exotic names such as neutrinos, WIMPS, or MACHOS); and 73 percent dark energy (see figures 7.6 and 7.7).

However, the situation is a little more complicated than this. It turns out that Sn 1997ff is actually *brighter* than expected and not fainter, and hence is closer than it should be. What this tells us is that initially the Universe *was* slowed by gravity, but that after a certain amount of time the acceleration took over progressively until now it dominates gravity completely. Sn 1997ff exploded about 2.4 Gyr after the Big Bang, when the Universe had 15 to 20 percent of its present age and thus was considerably denser and more concentrated than it is now. During this phase, gravity was still the dominant force; it was only when the density dropped further that the dark energy took over and provoked the acceleration.

One consequence is that Einstein was wrong to have said that the antigravity force he called the Lambda term was his greatest ever blunder. Having been convinced by Hubble's work that the observed expansion of the Universe made the

Lambda term unnecessary and incorrect, the Universe has had the last laugh. Einstein's Lambda term really exists and is extremely important.

The End of the Universe

With all we have found out, what will be the final fate of the Universe? Will it go out in a blaze of glory? Or will it suffer a long drawn-out agony? Given the information that we have now, even if it were not for the acceleration of the expansion of the Universe, it is well short of the amount of mass necessary to stop the expansion. Unless some new surprise awaits us to change everything once again, which seems somewhat unlikely, the Universe will expand forever at an ever-increasing rate.

What future then awaits the Universe? At present, galaxies like the Milky Way still have large amounts of fresh gas and are still forming stars. The rate of star formation will reduce progressively with time until finally all the gas has been used up. When that happens, the most massive and luminous stars will disappear rapidly. Smaller stars, like the Sun, will survive longer, but all the stars more massive and luminous than the Sun will disappear. About 10 Gyr after the last stars form, those that are the same mass as the Sun will turn into red giants and finally collapse into white dwarfs and die. From then on the lights will go out progressively in the Universe. Smaller stars will last for much longer than our Sun, but the most luminous stars that remain will become dimmer and dimmer dwarfs. A star with a quarter of the mass of the Sun will last 20 times longer than the Sun—about 200 Gyr—but will be a tiny glowworm compared to the Sun with around one-fiftieth of its luminosity. These, though, will, at that time, be the most luminous stars, the searchlights of the Universe. A consequence is that other galaxies will become faint objects. If our local group survives and the distances between the galaxies remain about the same as now, the Andromeda Galaxy would be a difficult object to detect, being thousands of times fainter than now and spread over a large area of the sky. More distant galaxies would become all but undetectable. The night sky would dim tremendously with the almost 6,000 stars currently visible to the naked eye being reduced to a mere handful; at times, it might well be that no stars would be visible in the sky because the movement of stars around our Galaxy is likely to give gaps of millions of years when there is no star close enough to our (sunless) Earth to be visible to the naked eye.

Things will get slowly but steadily worse from there. The smallest and dimmest red dwarf stars will last for approximately 1,000 Gyr but may be only a ten-thousandth of the luminosity of the Sun. These will be the last stars to die, but they will die in a dark and lonely Universe in which rarely, in our region of the Milky Way, will two stars be close enough together that one star will be visible from another with the naked eye.

As time passes, these tiny, dim stars will switch off one by one, leaving a Universe that will be cold and dark and filled with the dead shells of the stars that had once existed.

> *This is the way the world ends*
> *This is the way the world ends*
> *This is the way the world ends*
> *Not with a bang but a whimper.*
> —T. S. Eliot, "The Hollow Men"

SUGGESTIONS FOR FURTHER READING

Popular Books

Nigel Calder, *Violent Universe: An Eyewitness Account of the New Astronomy* (London: Penguin Books, 1977).

> *This is the paperback version of the superb book originally produced in 1969 to accompany the BBC superdocumentary of the same name (one of a series produced by Nigel Calder for the BBC in the 1960s and 1970s to explain the revolution going on in different fields of science—astronomy, geology, meteorology, nuclear physics, etc.—and the changes that these new discoveries were causing to our knowledge of the Earth and the Universe). Although now largely of historical interest, it is still quite easy to obtain (something that attests to its enduring quality) and well worth reading for the background it offers to the huge changes in astronomy in the 1960s that the discovery of quasars, pulsars, neutron stars, and the cosmic microwave background were bringing.*

Bernard Lovell, *Out of the Zenith: Jodrell Bank, 1957–70* (Oxford: Oxford University Press, 1973).

> *In this more advanced book, with some mathematics and theory, there is also a lot of exceptionally good narrative for readers who wish to gloss over the occasionally more difficult bits. Sir Bernard Lovell writes about the history and work of the Jodrell Bank telescopes and gives us a ringside seat into the story of*

how *Jodrell Bank and other instruments were used to try to distinguish the fate of the Universe. A fascinating story, beautifully told.*

More Advanced Reading

W. Baade and R. Minkowski, "Identification of the Radio Sources in Cassiopeia, Cygnus A, and Puppis A," *Astrophysical Journal* 119 (1953): 206–14.

This is the original article by Walter Baade and Robert Minkowski that identified the radio source Cygnus A as a peculiar kind of remote galaxy. The authors believed Cygnus A to be two galaxies in collision, an interpretation that was later found to be incorrect, but one that was believed for many years. The article is fascinating and is not particularly difficult to read. It can be found on the Internet at the following address as scanned GIF files: http://articles.adsabs.harvard.edu/cgi-bin/nph-iarticle_query? 1954ApJ . . . 119..206B.

A. Riess et al., "Observational Evidence from Supernovae for an Accelerating Universe and a Cosmological Constant," *Astronomical Journal* 116 (1998): 1009–38.

This article is most definitely not an easy read, but it is a historical article in that it was the first study to present observation proof that the Universe is inflating. The article can be read online at http:// cfa-www.harvard.edu/cfa/oir/Research/supernova/publications.html.

On the Internet:

High-Z Supernova Search Project
http://cfa-www.harvard.edu/cfa/oir/Research/supernova/home.html

Though not exactly for the beginner, this is a useful reference. It has a section for the public with images and explanations, plus detailed results for the advanced amateur or student.

KAIT
http://astron.berkeley.edu/~bait/kait.html

The Home Page of the Katzman Automatic Imaging Telescope used by the Lick Observatory Supernova Search contains details of the supernovae discovered by the telescope, many images of supernovae (as well as the telescope and other objects), news, and explanations.

How Will the Universe End?

CHAPTER 8

Why Is the Sky Dark at Night?

It seems odd to ask a question whose answer appears obvious, but the reason for a dark sky has been discussed for hundreds of years. It contains a lot of complex physics and cosmology, and the answer is in no way trivial. The observation that the night sky is dark gives important information about our Universe. Were it not for a particularly important piece of physics, life on Earth or anywhere else in the Universe for that matter, would be totally impossible.

What we now know as Olbers paradox has been discussed since at least 1744, although some experts attribute the first discussion of it to Johannes Kepler. The problem was first stated by Jean Philippe Loys de Chéseaux, a Swiss astronomer who lived in Lausanne, in 1744: "Why is the sky dark? If the number of stars is infinite, a stellar disk should cover every patch of sky."

De Chéseaux was a Swiss astronomer and mathematician who lived from 1718 to 1751. His name is best known because of his discovery and observations of the famous comet of December 1743 to March 1744, which became very brilliant and showed six tails in a famous peacock's fan.[1] Although de Chéseaux saw the comet on December 13, 1743, he was not actually the discoverer, even though the comet is known by his name. The Dutch astronomer Dirk Klinkenburg had seen it four days earlier, and the comet may well have been observed in November, too. This comet must be rated as one of the most beautiful and also one of the intrinsically brightest comets that has ever been seen. The comet was observed by 13-year-old Charles Messier, leading to a curious connection between the two observers. Messier, despite being the first great comet hunter, has always been best known for his catalog of nebulae: objects such as galaxies and clusters of stars that could be confused with a comet through a telescope.

In 1746 de Chéseaux compiled a catalog of 21 nebulae and star clusters that was presented to the French Academy of Sciences, of which no less than half a dozen were not later recorded by Messier.

De Chésaeux's question was a very good one. In 1744 astronomers were still a long way from discovering the true scale of the Universe. It would not be for almost another century, in 1838, that Friedrich Bessel would announce for the first time the distance to a star, but the telescope was already showing astronomers that, however far they looked into space, there were more and more stars to be seen. Nearly 200 years would pass until astronomers could measure the sizes of stars directly, and even if they had known the true size of the Sun, the sizes of other stars would not be known until well into the twentieth century. Nobody in the nineteenth century could know if the Sun was larger or smaller than other stars. However, Bessel reasoned correctly that such details were irrelevant.

It does not matter how large individual stars are, or how far away they are: if the Universe is infinite, as we look into space, more and more stars will appear to fill in the gaps between the brighter, nearer stars until the whole sky is filled with the blazing light of an infinite number of stars. This is similar to the grains of sand on a beach. Each grain is tiny compared to the size of the beach, but as billions and billions of grains join together, the beach is completely covered with sand, with more grains filling in the gaps until there is no space uncovered.

Given that the surface of the Sun is hot, in an infinite Universe the whole sky should be as bright and as hot as the surface of the Sun. In other words, life on Earth should be impossible because the whole sky would blaze with light at 6,000°C, cooking the surface of our planet.

Clearly, though, the night sky is dark. Where is the error?

The Paradox Redefined

In 1826 Heinrich Wilhelm Matthäus Olbers (1758–1840), the discoverer of the minor planets Pallas and Vesta, reformulated the paradox of why the sky is dark at night, making the situation even worse. The intensity of light reduces with the square of the distance from the observer. If the distribution of stars is uniform in space, then the number of stars at a particular distance from the observer should be proportional to the surface area of a sphere whose radius is that distance. At each radius, therefore, the amount of light should be both proportional to the radius squared and inversely proportional to the radius squared. These two effects will cancel, and so every shell should add the same amount of light. In an infinite universe, the sky would be infinitely bright.

In other words, according to the reasoning of Olbers, the sky would not even be at the temperature of the Sun: it would reach an infinite temperature.

Olbers was a remarkable astronomer whose many contributions are often not given as much credit as might be due to them. Born near Bremen, he studied medicine at Göttingen and at the same time attended a course on mathematics. While still at university he developed a new method of calculating the orbits of comets. This was later published along with a catalog of 87 comet orbits that was progressively enlarged by other astronomers who used his method. He also discovered three comets himself. Despite his great contributions to astronomy, Olbers was never a professional astronomer and practiced medicine for 40 years before retiring at the advanced age of 64.

To examine the question that Olbers formulated, let's reverse it and start with what looks like an even more trivial problem.

Why Is the Sky Bright during the Day?

This looks ridiculous. The sky is bright during the day because the Sun is up! Or is it? Think about this a little. If you go into space on the Space Shuttle, to the Moon, the sky is most definitely not bright even during the day.

In images taken by astronauts on the lunar surface, the sky is totally black, without the slightest trace of color or brightness. A telescope would be able to observe stars perfectly by day on the lunar surface, although the astronauts did not see stars themselves. The astronauts could not see stars because they were in a brilliantly lit landscape with little shadow, so the contrast with the bright surface meant that they were too dazzled to see stars. Had they gone into the shadow of the lunar module and lifted their gold-plated protective visors, they would have been able to adapt to the dark and see the stars far more brilliantly than in any earthly sky. Similarly, when they took photographs, the exposures were short ones with slow film, to show the surface and the Lunar Module, not to pick up stars in the background (see figure 8.1).

Why is the sky dark black on the Moon but bright blue on Earth? The answer is our atmosphere. The Earth's atmosphere scatters and disperses the light from the Sun. We see a blue color because the molecules in the air disperse the shorter wavelengths—the blue light—more than the red. The more atmosphere there is, the more strongly it disperses light. If you watch the sky, you will realize that the color is not constant. Meteorologists have used the color of the sky to predict the weather in a simple but effective manner for many years. Some days the sky has the very light blue color that you associate with fine weather. The light blue color indicates there is a lot of atmospheric dispersion, and thus high pressure. High pressure is normally associated with good weather. In contrast, a deep blue sky indicates that there is less atmospheric dispersion and thus low pressure. Amateur meteorologists know that a dark blue sky often precedes bad weather, which is usually brought by a low-pressure system—a depression.

The brightness of the sky thus functions as a quite effective barometer. As we rise in the atmosphere, for example, in a high-flying aircraft, the sky gets darker blue. Astronauts see the sky pass from light blue to dark, then purple and black as they rise into space and the atmospheric pressure drops with altitude. On Venus, with its thick atmosphere, the sky is actually red because the blue light is completely absorbed and the part that gets through the clouds to be dispersed is the red light.[2]

By the same token, the night sky is also not completely dark. Even when there is no Moon, light is dispersed. If you take a long-exposure photograph, the sky comes out the same shade of blue as during the day, and the stars are superimposed on this improbable color. We do not appreciate the blueness of the night sky because it is faint and our eyes are not sensitive enough to detect it, but a photograph can.

The Big Bang

Nowadays almost all astronomers agree that the Universe started with the Big Bang. For many years two theories competed for support: the Big Bang and the steady state theory.

The Big Bang theory stated that the Universe began with all the material of the Universe in a single point in time and space. This is often called the cosmic egg. At some point in time this cosmic egg exploded, "hatched" you might say, blasting into space all the material that now forms all the stars and all the galaxies. This material has continued expanding ever since, although the expansion slows gradually with time as the gravitational pull of the mass of the Universe pulls the material back. The alternative theory—the steady state—does not deny that the Universe is expanding. What it states is that the Universe never had a beginning and will never have an end, that material is constantly forming out of the vacuum of space to fill in the gaps opened by the expansion. The new material is in the form of hydrogen atoms and the rate of creation required is so slow—after all, most of the Universe is empty space—that it could never be detected.

Emotionally, the steady state theory was attractive, if for no other reason than it avoided the need to explain how the Universe started and how it would end. In the 1950s and early 1960s the two theories competed fiercely, each strongly supported by eminent astronomers. By the late 1960s, though, the balance swung heavily in favor of the Big Bang theory as it overcame different tests that astronomers proposed to check its validity. As we will see later, there was one observation that was the killer blow to the steady state theory and that has led to the Big Bang being almost totally accepted by most astronomers.

The Big Bang theory came in four varieties, each with its supporters.

1. The open universe: The Universe will expand forever. There is not enough mass to stop the expansion. As time goes on, the stars will die, ending with the small, faint, very long-lived red dwarf stars. Finally, the Universe will end cold and dark as the very last of the stars dies.
2. The critical universe: The Universe will continue expanding, but the expansion will slow progressively until it stops after an infinite amount of time. The result will be the same as in the open universe.
3. The closed universe: The Universe will expand at an increasingly slow rate

until it stops and starts to contract. This will happen if the mass of the Universe exceeds a critical value. The expansion will then be inverted with an ever more rapid contraction until all the mass of the Universe returns to its point of origin—the Big Crunch.

4. The oscillating universe: The expansion will stop, and the Universe will contract to give a big crunch. After an indefinite amount of time there will be a new Big Bang and a new cycle of expansion, contraction, and Big Crunch will start. In this case the cycle may repeat an infinite number of times, but we can never know how many cycles have gone before.

At present, all the evidence favors an open universe, but it continues to be the topic of animated and heated debate.

Six Popular Theories

Six theories have been proposed to explain why Olbers paradox is satisfied: in other words, why the sky is not a fiery inferno. Let us take a brief look at them before examining each one in more detail.

• *Dust stops us from seeing the distant stars.*

Clouds of dust like those found in our Galaxy block the light from distant stars. Only the light from more nearby stars reaches us, while the more distant ones that would fill the gaps are invisible.

• *The Universe has only a finite number of stars.*

Olbers paradox assumes that there is an effectively infinite number of stars in the Universe. If the true number of stars is small enough, then they will not superimpose and the gaps between them will ensure that the night sky is less bright than the surface of a star. The smaller the number of stars, the larger the gaps between them and the darker the sky.

• *Stars have only a finite lifetime.*

As stars have only a limited amount of mass, they can only live for a limited time. If the more distant stars have died by now, then the amount of light from further parts of the Universe will reduce sharply, and the paradox will be resolved.

• *The distribution of stars is not uniform.*

Even if there is an infinite number of stars, they may not be evenly distributed around the sky. If there are more stars in some directions than in others, there will be gaps in the sky in some directions, and in others the nearer stars would hide the more distant ones. The result is that not all the sky would be covered by stars.

• *The Universe is expanding.*

In an expanding Universe we would see the nearby stars as hot and as bright as our Sun. Distant stars would appear progressively fainter and cooler because of their red shift, which shifts their light further and further toward the red end of the spectrum and spreads it out further.

• *The Universe is young.*

If the Universe is not infinite, then there will not be enough distant stars to fill the gaps between the nearby ones. In a very young Universe there will simply not be enough distant stars, so that distant light has not reached us yet.

The Role of the Dust

We know that the Universe is full of cold dust and gas that blocks the light of more distant stars. You can see this for yourself if you go outside on a clear night in the Northern Hemisphere summer at some location where the sky is dark and transparent without light pollution. If you look toward the constellations of Scorpio, Sagittarius, and Cygnus, you will see that the pale band of the Milky Way is broken and torn by numerous pools of darkness (see figure 8.2). If you live in the Southern Hemisphere, you can see perhaps the darkest and most spectacular of these in the constellation of Crux. The Coal Sack, as it is known, is a region of the Milky Way where almost all the light from more distant stars is blocked by a particularly dense cloud of dust.

To astronomers, this dust, which is mainly concentrated in the plane of spiral galaxies like our own, is a severe nuisance. In certain directions our line of sight is rapidly completely blocked by clouds of dust. Close to the plane of the Milky Way we see very few distant galaxies. This is not because the galaxies are not there, but rather because dust completely blocks our view out of the Milky Way into intergalactic space.

Although de Chéseaux did not know about dust in the Galaxy—some fifty years would pass before Sir William Herschel produced the first rudimentary picture of the distribution of stars in the Galaxy that would, unknown to him, show proof of great dust clouds—he claimed that a slight loss of light from distant stars would solve the problem. His "loss of light" would much later be explained as due to the presence of dust. Astronomers later also proposed other much less enduring ideas—such as the theory that light, traveling for billions of years, might tire slightly on its journey and lose energy, getting redder and thus cooler in the process. The tired-light theory was popular in the first half of the twentieth century before it was shown to be incorrect. Surely, then, we need look no further than this most simple explanation. If dust is so ubiquitous in the Universe, then that would be the most obvious reason why the sky is dark at night: the light from distant stars is blocked.

This explanation is wrong because it contains a simple physical fallacy. Although the dust absorbs the energy, it does not go away. In other words, the dust blocking the heat from deep space would get progressively hotter and hotter as it absorbs energy until it, too, would be as hot as the surface of a star.

Thus dust will only buy us a period of grace before we inevitably fry from the combined heat of untold trillions of stars.

There Are a Lot of Stars, but That Is Not the Same as Infinity

Obviously, unless the Universe is infinite there cannot be an infinite number of stars in it. As the number of stars in the Universe, large as it is, is not infinite, these stars do not cover the whole sky. This is plausible, at least at first sight.

But how many stars are there in the Universe? In July 2003 a team of Australian astronomers announced the results of a census of all the galaxies in a region of the sky. From that, they estimated the number of stars in the Universe was 70 sextillion. If you prefer this as a number, it is 70,000,000,000,000,000,000,000 (7 followed by 22 zeros). Is this close enough to infinity so as to make no difference? The answer is yes. Seventy sextillion is enough stars to cover the whole sky.

The number of 70 sextillion is the best guess that has been made to date of the number of stars. It is more than the total number of grains of sand in all the Earth's beaches and deserts, but it only includes the stars in the visible Universe within range of our telescopes. When announcing the result, a member of the team that carried out the count, Dr. Simon Driver of the Australian National University, said the actual total could still be very much larger than this number. This means that we can be certain that there is no error and that the number of stars really is large enough to cover the whole sky.

More evidence that there are stars all over the sky was found in the 1990s by taking deep exposures of seemingly blank star fields, using the 2.5-meter Isaac Newton Telescope in La Palma (Canary Islands, Spain). When very long exposures (tens of hours) were taken, the whole of the field was filled with faint galaxies: there was no blank sky at all.

The Mortality of the Stars

As every star contains only a finite amount of matter and therefore shines only for a finite period of time, after which it runs out of fuel, we could resolve the paradox this way. De Chéseaux first proposed this theory in a slightly different form. He suggested that rather than stars dying, they switched on at a certain distance, although it would be more than 200 years before astronomers would understand how stars produce their energy and could explain how a star is born, and even longer before the processes of stellar death could be understood (in fact, they are not even fully understood now). The theory was first suggested in this modern form by the poet and writer Edgar Allan Poe. Could this be the resolution of the paradox?

We know that a star like the Sun will live for about 10 billion years. At present the Sun has existed for about 5 billion years, so it is middle-aged. We know that the Universe is about 12 to 15 billion years old. Recent results from the WMAP (Wilkinson Microwave Anisotropy Probe) satellite, launched by NASA to measure the tiny differences in the temperature of the microwave radiation from the Big Bang (see figure 5.5), suggest that the true age of the Universe is almost exactly in the middle of this range, being 13.7 ± 0.1 billion years. In other words, the Sun came into existence when the Universe was about a third younger than it is now.

The Sun is a medium-sized star. Many stars are smaller than the Sun, although most of the stars we observe are larger. One would assume that the larger the star, the longer its nuclear fuel would last. However, the situation is counterintuitive. Small stars survive for longer than larger ones because they use their nuclear fuel much more slowly. The most massive stars of all, perhaps 100 times the mass of the Sun, may live for only around 1 million years. In contrast, the least massive stars, with just a tenth of the mass of the Sun or less, may survive for hundreds of billions of years.

The oldest stars we observe in the Universe are around 13 billion years old. In fact, for a time they were responsible for a major scandal in astrophysics because these oldest stars appeared to be older than some estimates of the age of the Universe itself.

These facts point out the fallacy in the reasoning of Poe (and de Chéseaux). Stars are dying constantly in the Universe, but at the same time new stars are born and so replace them. The paradox stands if stars are constantly being born randomly across the infinite Universe (as we know they are), shine for a finite period, and finally die to be replaced by others.

Stars Are Not Evenly Distributed

For many years astronomers believed that what was termed the Cosmological Principle would be exactly fulfilled. This theory was later expanded to the exact fulfillment of the Perfect Cosmological Principle. What are these two principles?

According to the Cosmological Principle, the Universe has exactly the same appearance on a large scale from any point in space. It does not matter where the observer is in the Universe. It will be impossible from observing the sky to deduce where you are because the Universe is homogeneous and isotropic. Because all points appear to be the center, there is no "center of the Universe."

The Perfect Cosmological Principle is the same as the Cosmological Principle, but it adds that this state of affairs is not just true now but will be true at any time in the history of the Universe.

For many years it was believed that the Cosmological Principle is true, and many scientists believed that the Perfect Cosmological Principle would also be true. This rules out the possibility that stars are unevenly distributed in space to resolve Olbers paradox. Over the past 10 years, though, it has become obvious that the Cosmological Principle is *not* true.

If we look at the distribution of galaxies around the sky, there is growing evidence that there is structure in it. There are increasing suggestions of an apparent honeycomb structure. In some directions we see many more galaxies than in others: these seem to follow filamentary structures that have been termed *walls*. The most spectacular of these has been named The Great Wall by astrophysicists. In other directions there are deficits of galaxies (voids), the most famous of which is the Boötes Void, named after the constellation in which it is found. In figure 8.3, we see these structures as bright and dark regions respectively.

Everything suggests that galaxies and thus stars are not evenly distributed around the Universe. Models of the Universe that start with a rapid increase in size, termed *inflation*, suggest that the Universe should have a structure something like foam, with bubbles in the distribution of matter. The walls that we see would be the bubbles themselves, and the voids their interiors.

How does this affect our understanding of Olbers paradox? We do not yet know. A Universe with unevenly distributed stars would lead to the brightness of the sky being blotchy. Some regions of the sky would be bright, and others would be dark. A great deal of work is being done to understand the distribution of galaxies, but at present we do not know enough about their large-scale structure. The large map shown in figure 8.3 is drawn with just 2 million galaxies and still deals with relatively nearby structure. There are around 100,000 times more galaxies in the Universe than shown in this image, and it will probably be some years before large-enough scale maps are available to give definitive answers. It is now, though, certain that some structures are genuine.

The Expanding Universe

When Olbers reformulated the paradox, astronomers had no idea even of the distance to the nearest stars. There was real no concept of our Milky Way as a galaxy of stars even after Herschel published his map of the Galaxy in 1786. The first estimate of the size of the Galaxy was not made until Harlow Shapley's study

Cosmological Enigmas

was published in 1918. Herschel was the first to suspect that some of the nebulae that could be observed with his telescope were starry in nature and could be distant galaxies, but this was not proved until well into the twentieth century.

In the 1870s Lord Rosse had discovered that there were radical differences between different kinds of nebulae. Rosse had used his 1.8-meter (72-inch) telescope at Parsonstown in Ireland to examine many nebulae, finding that a number of them could be resolved into stars. These nebulae broadly divided into "gray" nebulae, some of which he was astonished to find were spiral in form and resolved into stars—these were later shown to be external galaxies—and the "green" nebulae that we now know to be gas clouds in our Galaxy.[3] Even in 1890, however, after the first spectroscopic observations of nebulae, there was still huge uncertainty about their nature. Writing in 1890, J. E. Gore speaks of different classes of nebulae as irregular nebulae (what we now call nebulae, or gas and dust clouds); elliptical nebulae (here, galaxies such as the Andromeda Nebula are mixed with globular clusters such as Omega Centauri and even planetary nebulae such as the Dumbell, in the constellation of Vulpecula); annular nebulae (now known as planetary nebulae); spiral nebulae (spiral galaxies); and planetary nebulae. Note how three totally different types of object were all lumped together as "elliptical nebulae."[4]

Even as late as 1920 there was a famous debate between Harlow Shapley and Heber Curtis about the nature of the Milky Way and the spiral nebulae. Curtis held the view that the solar system is at the center of the Milky Way but, in contrast, accepted that the spiral nebulae were "Island Universes" well outside the Milky Way. In contrast, Shapley argued that the solar system lies some 15,000 parsecs from the center of the Galaxy, but believed that the spiral nebulae, such as the Whirlpool galaxy M51, were gas clouds within the confines of the Milky Way. Both astronomers were right in one of their postures and wrong in the other, but it is astonishing to think that even an astronomer of the status of Shapley could still not accept the existence of external galaxies.

By then, the discoveries of an American astronomer, Vesto Slipher, whom we have already met in a completely different context, that of the search for Pluto,[5] had already sounded the death knell of the galactic isolationists. In 1912 Slipher had used the 60-centimeter (24-inch) Clarke refractor at Lowell Observatory in Flagstaff, Arizona, to take spectra of galaxies. Working with photographic plates of low sensitivity on such a small telescope meant that some of the galaxies that he observed required as much as 80 hours of exposure time, which makes Vesto

Slipher's work even more extraordinary. All but two of the galaxies he observed showed a red shift; in other words, they were receding from the Earth, with the largest velocities that he could measure being around 1,000 kilometers per second. Had Slipher had access to a slightly larger telescope, he would have had the opportunity to make an even more important discovery that, in the end, fell to Edwin Hubble in 1923. Hubble had access to the 2.5-meter (100-inch) Hooker reflector at Mount Wilson Observatory and used it to measure the distance to the Andromeda Nebula, M31. Although the value that Hubble measured was far too small—900,000 light years, when the true value is 2.2 million—there was now no doubt that it was a true external galaxy comparable in size with the Milky Way.

By 1929 Hubble was able to publish his discovery that the more distant the galaxy, the larger its red shift, and thus to prove that the Universe was expanding.

Curiously, a few years earlier Albert Einstein had discovered that his equations predicted that the Universe would be in expansion. Unable to understand this result, he added his "fudge factor" to his equations to counteract the expansion. Einstein called this the cosmological constant and later lamented that to use it instead of recognizing the reality of the expansion of the Universe was his greatest mistake.[6]

The expansion of the Universe has two important effects:

1. More distant objects appear redder and their light thus less energetic.
2. The light is more spread out. You can compare this to a balloon that is inflated: the same amount of rubber is spread out over a much larger area. In the spectrum the light is spread out over a wider range of wavelengths.

In many places you will read that the expansion of the Universe is the explanation of Olbers paradox and that the observation that the night sky is dark proves that the Universe is expanding. Certainly the fact that the more distant galaxies are effectively dimmed by the red shift does contribute significantly to the resolution of Olbers paradox, but it is not the major contribution.

What is the major reason why the sky is dark at night?

Cosmological Enigmas

The Universe Is Young

Olbers paradox supposes that the Universe is effectively infinite in extension. We have already seen that the number of stars in the Universe is very large, but not infinite, although for the purposes of the paradox it is sufficiently large so as to make no difference to the result.

The paradox is resolvable in a variety of ways. If the Universe has existed for only a finite amount of time, as the Big Bang theory states, then only the light of finitely many stars has had a chance to reach us yet, and the paradox breaks down. If the Universe is expanding and distant stars are receding from us (also predicted by the Big Bang theory), then their light is red-shifted. This diminishes their brightness, again resolving the paradox. Either effect alone could resolve the paradox, but both are working together, and calculations show that the finiteness of time is the more important effect.

We have seen that the Universe appears, according to the latest measures, to be 13.7 billion years old. It is not infinite in age and is far from infinite in extension, and these are the most important reasons why the sky is dark at night. Were the Universe larger and older, then the light from more and more distant stars would pile up and give us a bright and furnace-like sky at night.

However, having argued that the sky is dark at night, we have one more discovery to make. The Big Bang has a surprise for us. The sky is *not* dark: the sky is bright at night; it is just that we do not have the right kind of eyes to see it.

The Light at the Edge of the Universe; or, The Problem with Pigeons

The discovery that convinced most astronomers that the Big Bang model of the Universe is correct was made quite by accident, in a way similar to the discovery of radio waves from the center of the Milky Way, which had been made accidentally by Karl Jansky in 1927. What is less well known, though, is that the discovery was snatched from under the noses of a renowned group of astronomers who were trying to make the discovery for themselves.

By 1965 a young radio astronomer called Arno Penzias had been testing a new kind of radio antenna at Bell Laboratories at Holmdel in New Jersey for several years. The new antenna, in the form of a horn six meters across, had been developed by AT&T with the aim of using a chain of similar antennas to relay long-distance telephone calls (today, microwave links are extensively used for this purpose). This antenna was causing considerable frustration at Holmdel. Arno Penzias was working with another astronomer, Bob Wilson, to try to detect microwaves from the Milky Way using the antenna. Unfortunately, just as Karl Jansky had been in 1927, Penzias and Wilson were being driven to distraction by a faint hiss in the antenna that they could not get rid of and that frustratingly came from all over the sky. In desperation they put the blame on pigeon droppings in the antenna and went through the process of cleaning the inside of the antenna thoroughly. To their horror, even after they cleaned the antenna, the hiss was still there.

At this point, chance took a hand. At MIT Bob Dicke had suggested that it might be possible to detect very cool radiation from the edge of the Universe; he realized that by improving some existing equipment he might be able to detect microwaves from the Big Bang. Even though the Big Bang itself was inconceivably hot, as the explosion expanded, it cooled, and it was estimated that radiation should still be detectable, with the temperature now just a few degrees above absolute zero. The improved equipment was built and installed on the roof. In the meantime, though, a colleague of Dicke gave a seminar on the experiment and its implications. One scientist in the audience commented on the seminar to a friend. That friend told another friend about it. And that friend was a colleague of Penzias. Reportedly Penzias then telephoned Dicke, explained that they were detecting a faint microwave signal that appeared to come from all over the sky, and asked if it might not be from the Big Bang; when Dicke heard this, he realized that Penzias had found the radiation that his team had hoped to detect. When he put down the receiver, Dicke just lamented: "We've been scooped!"

It did not take long for the signal to be confirmed and its temperature to be measured. The value of 2.7K (−270.4°C) was very close to the expected value if the Big Bang had been cooling for about 10 to 15 billion years. Penzias and Wilson won the Nobel Prize in 1978 for their discovery.[7] Dicke won nothing.

This radiation is now known as the Cosmic Microwave Background (CMB), and its study is still one of the most important topics in astronomy. Over the years it has been found that it is a perfect black body—in other words, it emits absolutely perfectly evenly, following the theoretical curve of emission, as an object at 2.7K should. This in itself is a fundamental measurement. Were it to be found

that at some wavelength the CMB emits more strongly than expected, the Big Bang model of the Universe would be in serious and probably fatal trouble. However, in 1989 the COBE (Cosmic Background Explorer) satellite showed that over a quite wide range of wavelengths the CMB is a perfect black body to an astonishing 0.01 percent.

The CMB has another surprising characteristic. It is incredibly even over the entire sky. Great effort has been put into searching for unevenness—differences in temperature from point to point in the sky. With one exception, these differences are tiny, a few hundred thousandths of a degree centigrade at greatest. The exception is toward the constellation of Virgo. There the CMB is a fraction warmer. We know that there is a huge supercluster of galaxies in that direction and the slight additional warmth is caused by the Milky Way and our Local Group of galaxies falling toward the Virgo supercluster, attracted by its phenomenal pull of gravity and thus blue-shifting slightly the light of the CMB.

This smoothness of the CMB has caused no end of problems for cosmologists because it implies that the Big Bang itself was astonishingly symmetrical. In brief, it makes forming galaxies and stars very difficult. If the Big Bang were too even, there would be no irregularities in its density to serve as seeds to pull mass together to form galaxies. These irregularities, called anisotropies, were finally detected in 1991 and represent patches of the sky where the material was beginning to clump together to form the first protogalaxies perhaps a million years after the Big Bang.[8] The first anisotropy to be discovered was at a level of approximately 17 millionths of a degree centigrade.

The Bright Sky and the Universe

In other words, the sky is bright at night, but only to beings with eyes sensitive to microwaves (the same ones you use to heat your dinner at home). The brightness, distribution, and color of this light give an astonishing amount of information about our Universe. Not only do they confirm the Big Bang, but they allow us to detect the first galaxies starting to form soon after the Big Bang.

At the same time, the fact that—to our human eyes—the sky is dark at night also gives us a great deal of information about the Universe and tells us that it is both expanding and has a finite age.

It is astonishing that such simple observations can give rise to so many fundamental conclusions for cosmology.

SUGGESTIONS FOR FURTHER READING

Popular Books

Edward Harrison, *Darkness at Night: A Riddle of the Universe* (Cambridge, Mass.: Harvard University Press, 1987).

> *This book is now slightly dated by recent discoveries such as inflation, but is well worth reading still for its thorough review of the problem of Olbers paradox and its resolution.*

Michael D. Lemonick, *The Light at the End of the Universe: Dispatches from the Front Lines of Cosmology* (Princeton, N.J.: Princeton University Press, 1993).

> *This book covers a range of issues such as the mass and age of the Universe, dark matter, and the cosmic microwave background. The book is written by a science writer, not a professional scientist, and explains what is going on at the cutting edge of modern cosmology, who the personalities involved are, and how they think.*

More Advanced Reading

Paul S. Wesson, "Olbers's Paradox and the Spectral Intensity of the Extragalactic Background Light," *Astrophysical Journal* 367 (1991): 399–406.

> *A highly technical article that examines how the amount of light received from distant galaxies allows Olbers paradox to be resolved. The author calculates how the brightness of the sky would be affected by different suppositions about the expansion of the Universe and by the age of the galaxies, concluding that, contrary to the popular explanation that the expansion of the Universe is responsible, the main reason why the sky is dark at night is due to the finite age of the galaxies.*

Alan Guth, *The Inflationary Universe* (New York: Perseus Books, 1997).

> *Despite treating a highly technical subject and not being an easy read, this is a fascinating and well-written book. Alan Guth explains theories of cosmology and the origin of the Universe and their relation to modern discoveries in elementary particle physics. The book shows how the theory of inflation came about and the predictions that it makes about the Universe and its structure.*

Cosmological Enigmas

On the Internet

Yahoo directory of Olbers paradox sites

http://dir.yahoo.com/Science/Astronomy/Stars/Olbers__Paradox/

> *This site gives a listing of half a dozen good sites on the history and explanation of Olbers paradox. The sites range from very simple explanations of the problem to quite complex documents. Take your pick.*

CHAPTER 9

How Do We Know There Was a Big Bang?

At present just one theory of the Universe is widely accepted. In the late 1960s, three theories competed, each with a significant level of support. The current monopoly of scientific opinion for such a fundamental part of the whole basis of astronomy is unusual; normally there are several competing theories, even if one has a substantial majority of support in the scientific community. Nowadays, only a small rump of supporters of the oldest rival theory still attempt to hold back the scientific tide and, like King Canute before them, are being swamped by it.[1]

For many years the steady state, Big Bang, and oscillating universe theories stood side by side. Back when all the theories were still respectable, many hours were dedicated to discussing the different arguments. (In the Bristol Astronomical Society, in England, we staged our own great cosmological debate in the mid-1970s. After an at times highly amusing discussion, the oscillating universe theory won the day, unlike in the overall scientific community. Our musings, though, were of no consequence whatsoever in the overall debate.)

We have already explored the origins of the competing theories.

What we have not done is to look at the series of compelling discoveries that killed the steady state theory for all but its most unconditional supporters and that confirmed the basic tenets of the Big Bang theory.

Unsettling Times

The first half of the twentieth century was an unsettling time for astronomers. Einstein's theory of gravitation had greatly complicated the Universe. The discovery of the expansion of the Universe and the distances to the galaxies had, in turn, expanded our horizons more than 1,000-fold. And the "et tu Brute" was the suggestion that if stars lived as long as seemed possible if they converted mass directly to energy, then a star like the Sun should continue to shine for around 100 trillion years.[2] (Even though this estimate turned out to be far too long, the true value was still hundreds of times greater than was believed at the end of the nineteenth century.) The new discipline of cosmology—the study of the cosmos around us—was just being born. It is a measure of how disconcerting events were that all three of the aforementioned bastions that support cosmology were initially strongly resisted by some scientists. Two of the three—the expansion of the Universe and the theory of relativity—are *still* being resisted by a small minority.

So, although scientists had only measured the size of our own Galaxy for the first time in 1918 with the stunning result that it was much greater than previously thought, within five years they were having to face up to the fact that our Galaxy was a tiny part of an immensely vast universe and that our Universe was expanding and getting bigger to boot. Add in the fact that stars seemed to have lifetimes that were to all intents and purposes infinite and it is easy to see why astronomers were disconcerted.

Astronomers reacted to these vertiginous facts in one of two ways. For some, it was natural to think that, if the Universe was expanding, it had an origin in a definite point in time and at a fixed point in space. This view, though, led to some important philosophical questions that were extremely hard to confront. What was there before the Big Bang? What caused the cosmic egg to come into being in the first place? What then caused the cosmic egg to explode and give rise to the Universe? The basic answer to all three questions was the same in 1930 as it is now: we do not know! This led to a strong negative reaction from some scientists. If we cannot explain *why* the Universe came into being, do we actually *need* it ever to have come into being? In other words, do we need a seemingly infinite universe to have had a beginning?

This was how the steady state theory came into being. In essence what the steady state theory said was that the Universe has always existed and would always exist. What is more, it had always looked the same and would always look the same, being totally unchanging. In other words, the problem of explaining what happened before the Universe was created would be solved by the simple expedient of making the question meaningless—if there was no creation of the Universe, no beginning, there would be nothing to explain. End of problem!

The astronomers who formulated this theory in 1948 were no fly-by-night dreamers. Fred Hoyle, the motor of the group, Thomas Gold, and Hermann Bondi were top-flight theoreticians. During the Second World War Fred Hoyle had worked on radar countermeasures, but he was already developing a reputation as a theoretician, having worked at Cambridge under the supervision of the legendary Paul Dirac, before starting to work on stellar evolution. It was the theory of stellar evolution that was to cement his place as a great theoretician. First he worked on the theory of the evolution of red giant stars with Martin Schwarzschild, the son of Karl Schwarzschild. Then, in 1957 Fred Hoyle participated in one of the great scientific papers of the century—the so-called B^{2}FH paper—in which William Fowler and the husband-and-wife team of Geoff and Margaret Burbidge worked with him to explain the nuclear reactions that give rise in stars to elements heavier than helium; this paper was a fundamental plank in the understanding of stars.

Although the main thrust of his work had been into stellar evolution and how elements are formed in stars, Fred Hoyle had the sort of inquiring and restless mind that was always searching for new mysteries on which to work. He was struck by the famous solutions of the equations of the general theory of relativity in which the Universe was not static. The work of Vesto Slipher and Edwin Hubble had confirmed that the Universe was genuinely expanding as these solutions of the general theory of relativity predicted. An expanding Universe that does not renew itself in some way would be a decaying system, condemned to die. This struck Hoyle as an unsatisfactory situation. The idea that all the mass of the Universe had been compressed into a cosmic egg that, for unknown reasons, simply chose to explode, also struck him as being highly unsatisfactory.

He proposed the steady state theory in 1949 to address these issues. It rap-

idly won widespread approval from many cosmologists who welcomed a theory that permitted some order in the Universe and that offered a safe, simple alternative to the Big Bang.

It is less well known that Fred Hoyle was responsible for the name Big Bang. In 1949 he appeared on a BBC Radio program called *The Nature of Things*. During the program he defended his steady state theory for the Universe and, when asked for his views on the idea that the Universe had started at a definite moment and point in time in the past—the idea first suggested by Georges Lemaître in 1927 and later developed by George Gamow—Fred Hoyle responded that he could not conceive of the Universe as having been born in what he contemptuously referred to as a "Big Bang." Some months later, in 1950, the BBC published a summary of the broadcast in its magazine *The Listener*. As they would say on television, the explanation of Hoyle's name for the model—the Big Bang—is "not suitable for a family audience," and there is no possibility that my editor would let me slip it through even as a footnote. I will give a broad hint though: there is sexual innuendo involved, and the name means just about what you think it does. Given the standards of the time that applied to radio, television, theater, and the written word, it is astonishing that the BBC was prepared to risk publication of this phrase. Theaters were regularly being closed by the Lord Chamberlain's Office for promulgating unacceptable material—this could be any kind of sexual innuendo—with comedians who were prepared to push the limits, frequently causing the curtain to be brought down to stop their act rather than risk the Lord Chamberlain's wrath. This hugely influential publication was widely read, and ironically the term stuck. Little was Fred Hoyle to imagine that this throwaway comment was going to give the rival theory the name that was to impact so much on the public consciousness and contribute so much to its popularity.

The steady state theory of the Universe had one enormous problem—it violated the conservation of mass and required the amount of material in the Universe and hence its mass to increase continuously—but its simplicity and optimism made it popular. Late in the 1950s it is fair to say that even if the steady state theory did not ever quite muster majority support, it certainly had the support of a large minority of scientists.

It was not until the late 1960s that things started to go seriously wrong for the steady state theory.

The Lying Red Shift?

The great bastion and raison d'être of the Big Bang theory is the red shift. The fact that almost every galaxy in the Universe, with a few nearby exceptions, seems to be moving away from every other galaxy with a velocity that increased the further away the galaxy is from us—what is termed the cosmological red shift—seemed to set an open-and-shut case for an expanding Universe. The so-called cosmological red shift, however, has not been without its critics. In fact, it has suffered a series of major attacks over the years. Could it be that galaxies are not as far away as the red shift would indicate?

For a number of years, there was an alternative explanation for the red shift. What if light itself got tired, losing energy when traveling great distances through space and becoming progressively more reddened? Then, distant galaxies might not be anything like as distant as they seemed. In other words, the red shift would be lying to us, and only rather small red shifts would be Doppler shifts. The nearby galaxies would genuinely be moving away and at the distances calculated, but, as the red shift got larger and larger, an increasingly small part of the red shift would be due to movement and an increasingly large part due to light tiring. If this hypothesis were true, there would not be an expansion of the Universe on the scale suggested by the cosmologists, and hence the Big Bang would lose a large part of its need to exist.

There is still a tiny minority of cosmologists who propose the tired-light hypothesis as an alternative cosmology. Such models are termed *nonstandard cosmologies* as they defend precepts that are rejected by the vast majority of the scientific community. Although astronomers look on such models with a certain disdain, or even amusement, it is fair to say that astrophysics has a habit of occasionally showing that nonstandard models of the Universe are spectacularly right. Nicolas Copernicus was an example of someone who challenged all standard thinking and turned out to be correct. Before rejecting "standard thought," however, we should remember that for every Copernicus, there are a thousand dreamers who are spectacularly wrong. Recent discoveries in astrophysics have gone strongly against the predictions of the tired-light model. In particular, it predicts that the light curves of supernovae at high red shift should look essentially identical to those at low red shift. What is seen, though, is that supernovae at high red shift fade much more slowly than those at low red shift. The difference in the rate of fade between the high red shift and low red shift supernovae is exactly what is predicted by time dilation (that is, the slowing of time at velocities close to the speed of light). In other words, the observations of supernovae are exactly what is expected if the red shift is genuinely a Doppler shift.

The same arguments apply to a second and more modern warhorse. Since the early 1970s Halton Arp has presented a large number of cases of alignments and unusual groupings of galaxies with different red shifts.[3] Many different types of alignment have been suggested. There was much discussion about cases that were presented where several quasars with different red shifts *seem* to line up exactly with a nearby bright galaxy. The suggestion was that somehow these quasars had been shot out of the center of the galaxy and that their red shift was either due to the violence of their expulsion, or totally unrelated to the Doppler shift. The suggestion that quasars had been expelled violently from nearby galaxies could be dealt with easily: no quasars could be found that showed a blue shift in their spectrum and were thus coming toward us; unless every single quasar just happened to be expelled directly away from us (staggeringly unlikely), some of them *had* to show a blue shift. Similarly, astronomers looked in vain for any evidence of lateral motion; if a quasar had been expelled from a nearby galaxy at nearly the speed of light, it would be expected to move slightly in the sky, unless, of course, its movement was exactly away from us.[4]

A number of prominent astronomers insisted that *something* was wrong, however; despite the lack of blue shifts, there were far too many cases of alignments: more than would be expected if left to pure chance. The arguments about these cases became extremely statistical. In a nutshell, critics suggested that if you spread thousands of quasars and thousands of bright galaxies unevenly around the sky you are bound to get some chance alignments; the question was, were there genuinely more of these than expected or not? Careful analyses were published that suggested there were no more good alignments than would be expected by pure chance, even if you do not take into account the possibility that more quasars are likely to be discovered in the carefully studied regions around bright galaxies than in any other randomly chosen area of the sky.[5]

The evidence that the red shift is genuinely a Doppler shift, at least relatively close by in the Universe, was so indisputable, however, that not even the most ardent doubters were really prepared to challenge it. One alternative interpretation of red shift proposed by the skeptics taking this fact into account was that it had two components: one a genuine Doppler shift indicating distance, and another, of comparable size, generated by some other mechanism. These ideas have been supported strongly by scientists of the caliber of Geoffrey Burbidge.[6]

Most controversial of all have been the proposed cases of physical connections between quasars and galaxies. Of these, the most famous case is the pairing of the bright galaxy NGC 4319 and the quasar Markarian 205 (see figure 9.1). The quasar Markarian 205 apparently lies within the outer spiral arm of the barred spiral galaxy NGC 4319. The red shift of the galaxy (0.00468) corresponds to a distance of only 80 million light years, while the red shift of the quasar (0.071) is equivalent to a distance of approximately 1 billion light years. What interested Halton Arp was the fact that, although the quasar has a red shift suggesting that it is 13 times more distant than the bright and nearby galaxy, there appears to be a faint luminous bridge of material linking the two. If the red shift is telling the truth, it is impossible that galaxy and quasar can be physically connected.

For many years a controversy raged about this pairing. Critics argued that the luminous bridge was either an artifact of the processing of the images or due to the casual interpositioning of another galaxy between Markarian 205 and NGC 4319. Conspiracy theorists denounced mainstream science for not taking Arp seriously and suggested that their refusal to refute his arguments was proof of their fear of his views. It was even suggested there was pressure to stop scientists observing Markarian 205 with the Hubble Space Telescope in case they proved that Arp was right.

In 2002 Roger Knacke of Penn State Erie observed the pairing with the Hubble Space Telescope, producing some spectacular images. For most astronomers, previous spectroscopic observations taken with the telescope in 1991 had just about settled the case. Those observations showed the galaxy absorbed light from the quasar (in other words, there was a series of dark absorption lines in the spectrum of the quasar that were caused by light from the quasar being absorbed by cool gas in the galaxy's spiral arms). Knacke's images of the pair were the best ever, with a resolution many times better than the original ones published by Arp in 1971. The results were spectacular. Markarian 205 lies in a quite normal galaxy with a hint of spiral arms. The brightness of the galaxy is consistent with its distance calculated from the red shift. There is also a much fainter and less well known companion of Markarian 205 clearly shown in the Hubble images that is another com-

Cosmological Enigmas

pact but much fainter galaxy at the same red shift as Markarian 205. This forms a genuine pairing with only a small separation in space.

These images have not closed the issue of the luminous connection between quasar and galaxy; however, they have greatly reduced its importance. In the original photographic images published by Arp in 1971, the connection appeared to be quite clear and obvious. Despite the enormous resolution and detail in the Hubble images, this luminous bridge is seen less clearly than ever. Most scientists suspect it is a random wisp of structure in one or the other of the two galaxies, although there is no really clear explanation. One detail that makes one suspicious is that the lower the resolution of the images, the more clearly it is seen, which suggests that it is due more to some optical effect than to a real structure.

At the end of the second millennium most astronomers felt that the controversy over the cosmological red shift had really been closed. Although a tiny minority of scientists continues to express doubts, the general opinion that the red shift is cosmological—in other words, that it does indicate that quasars and, increasingly, faint galaxies are at enormous distances—seems to have overcome all reasonable doubts and tests. Why, though, was this argument over the distances of quasars so critical in the debate about the origin of the Universe?

The Long Agony of the Steady State Theory

Both the Big Bang and the steady state theory made a series of fundamental predictions about the Universe. For the steady state theory, the following predictions were the most important.

• *The value of the deceleration parameter* (q_0) *would be −1.*

The deceleration parameter expresses the rate at which the expansion of the Universe is slowing. It is often expressed in terms of the Greek capital letter omega (Ω) for the density of matter in the Universe as a fraction of the quantity needed to close the Universe, where $\Omega \equiv 2q_0$. In other words, if $\Omega = 1$, $q_0 = 0.5$, and the Universe is just closed, in the sense that there would be enough mass for its pull of gravity to stop the expansion after an infinite amount of time.

As the steady state theory, however, proposed that the mass of the Universe would always be increasing, albeit extremely slowly, maintaining its appearance unchanged with time would require a steady acceleration of the expansion at a constant and exactly defined rate to counteract the additional mass and thus to maintain the Universe "inflated." This led to the prediction that $q_0 = -1$.

We measure q_0 by looking at the brightness of distant objects and how they vary in brightness with distance. Suppose we observe 10 galaxies of the same luminosity that lie at different red shifts and thus distances. We should see them become steadily fainter in a regular manner as the red shift gets larger and thus the galaxy more distant. If the Universe were expanding at a constant rate with no deceleration—what is called an Empty, or Einstein–de Sitter Universe—the objects would lie on a neat straight line when we plot brightness against red shift. If the expansion of the Universe is *decelerating*, then distant objects at high red shift should appear brighter than expected because they will be closer to us than they would be if the expansion were not decelerating. In contrast, if there were an *acceleration* of the expansion, we would expect distant objects to appear fainter than expected for their red shift, because they will be further away than if the expansion rate were constant with distance.

From early on it was obvious that, as galaxies at higher and higher red shift were observed, the ones at large red shift were brighter than expected and their magnitudes did not follow the curve predicted by the steady state theory.

Of course, it was possible that nearby galaxies were not like the more distant ones. Possibly more distant galaxies were intrinsically more luminous than galaxies close to the Milky Way; hence they appeared to be brighter and thus closer than they really were. This, though, fell afoul of the second prediction of the steady state theory.

• *On a large-enough scale, the Universe would look the same at all times and in all directions.*

If small quantities of matter were being created constantly to fill up the void left behind by the expansion of the Universe—"small" in this context being around 100 atoms per year in the entire volume of the Milky Way—there would be constant renovation of galaxies. New stars would form out of the new material created as the old stars died out, and new galaxies would fill the voids left between the old ones. In other words, if we move far enough away from the Milky Way to get outside the local clusters and superclusters, the Universe would always look the same and would never change. This would lead to two additional predictions.

- *The average luminosity of galaxies would not change with time.*

- *The average number of galaxies in a given volume of space would not change with time.*

By the late 1960s both of these predictions were looking untenable. The plots of the brightness of galaxies against their red shift showed that the best fit to the deceleration parameter was the unrealistically large value of $q_0 = 1.6$; this would have implied that the mass of the Universe was more than three times the critical mass necessary to close it. With the best will in the world, this was far in excess of the largest amount of matter in the Universe that astronomers could imagine. Despite carefully selecting galaxies to be as similar as possible and thus presumably of the same luminosity, studies of some of the most distant galaxies suggested that q_0 might be as large as 12—although if the Universe truly were that massive, it would already have collapsed back on itself. The only tenable conclusion was that the luminosity of galaxies was not constant with time, which was just the opposite of what the steady state theory predicted.

More serious for the steady state model were the quasars—hence their importance in the cosmological battleground. By the end of the 1960s, the number of known quasars was in the hundreds and, by the end of the 1970s, was well past 1,000 and growing at a great rate. What was evident almost from the start was that the number of quasars changed rapidly with time. Think of a telescope as a time machine looking back into the past; the bigger and more powerful the telescope, the further back it can look in time. The larger the red shift, the more distant the object, and the further back in time we look. Conversely, at small red shifts we are looking at the recent history of the Universe.

There were no quasars at the extremely small red shifts that would place them in the local universe; the number increased rapidly with red shift and seemed to have a maximum at a red shift of approximately two, but at higher red shift the number of quasars dropped with extreme rapidity. By a red shift of four, all the quasars had disappeared. In other words, there seemed to be an "age of quasars," with no quasars either nearby—in other words, in the recent history of the Universe—and few, if any, extremely distant quasars—that is, in the early life of the Universe.

The fact that the number of quasars changes so evidently with time was totally counter to the predictions of the steady state model of the Universe. However, there was a possible way out; if the red shift of quasars was cosmological, if quasars were a lot closer and less luminous than they seemed, then their unusual distribution could be explained conveniently—all quasars would be nearby and genuinely distant quasars would be too faint to be observable. This was a large part of the motivation for the battle of the quasar red shifts that consumed much of the 1970s and 1980s. By the early 1990s the argument had largely died out, as discovery after discovery increased the conviction that quasars were genuinely at cosmological distances.[7]

Even if there were no other problems, the problem of the quasars would have been a mortal blow to the steady state theory, with the question of the value of the deceleration parameter no more than the "et tu Brute." Even before quasars became a serious problem, however, the question of the Cosmic Microwave Background (CMB) radiation had already put a rather large wrench in the works.

The Cosmic Microwave Background and Its Consequences

The existence of the CMB as a consequence of the cooling of the Big Bang was predicted in the 1940s by George Gamow, Ralph Alpher, and Robert Hermann, although not until 1964 was it was detected, albeit totally by accident, by Arno Penzias and Robert Woodrow Wilson. Calculations showed that the temperature of the CMB radiation was about 3 Kelvin (that is, 3 degrees above absolute zero, or −270°C), close to the value predicted if it were a relic of the Big Bang.

Many cosmologists viewed the discovery of the CMB as the final proof of the Big Bang, but then an old argument first proposed in 1941 by an astronomer named Andrew McKellar was rediscovered. McKellar had suggested that if the light of distant galaxies were scattered by dust and gas in the Universe, the background of space should have a temperature of about 2 Kelvin (2K). Because this was close enough to the observed temperature of the CMB, supporters of the steady state model could suggest that maybe the existence of this background radiation had nothing to do with the Big Bang.

The best that could be said in 1970 was that even if a majority of scientists felt that the CMB was a result of the Big Bang, the alternative explanation proposed by the steady state model could not be ruled out. Why is it then that we do not hear of this alternative explanation for the CMB today?

The answer is that in the 1970s experiments started to become sensitive enough to study the characteristics of the CMB—its exact temperature, its spectrum, and its brightness around the sky—in great detail. The results showed that the CMB was extremely pure; for one thing its brightness was exceptionally even around the sky, with no detectable differences; in addition, the temperature was pure in that the spectrum of the CMB came from a single, precisely defined temperature rather than a spread of temperatures. If the CMB came from scattered starlight, it would not be "pure"—stars have a wide range of temperatures, which would smear out the spectrum compared to radiation from a single temperature. It would be far less smooth around the sky, because different parts of the CMB around the sky would be of widely different brightness.

Ten years after the CMB was discovered the evidence was accumulating that the steady state theory could not explain it adequately, although the Big Bang theory probably could. Over the years more and more detailed studies of the CMB have eliminated alternative explanations, to the point where we can say that it is almost impossible to find a scientifically valid explanation for it other than the Big Bang.

Had the steady state theory suffered from a single one of the problems detailed here, it would probably have been fatal in the end. As it was, the steady state theory suffered from a steady hemorrhage of support due to the death by a thousand cuts. That, though, is not the same as proving that one of the alternative theories was correct.

Why Do Scientists Believe in the Big Bang?

The Big Bang theory has been successful because, up to now and unlike its rivals, it has met all the challenges that it has been faced with and made some stunningly successful predictions. In this it has also been aided by the fact that the other, rival theories have singularly failed to make successful predictions. Of course, it is possible that tomorrow, or next year, or in 10 years someone may make a discovery that cannot be accommodated within the Big Bang theory, and it will either have to adapt, or it will fall; at present, though, it does seem that the Big Bang has faced off all its possible rivals and that there are no major challenges on the horizon.

Let us look at the evidence for the Big Bang from the viewpoint of this theory.

The Expansion of the Universe

The Big Bang theory was a natural consequence of the discovery of the expansion of the Universe. If we accept the expansion of the Universe, unless we choose the somewhat artificial expedient of new matter appearing spontaneously to fill the void that forms between the galaxies and thus the possibility of a never-ending expansion, we are automatically stating that the expansion will have started from a certain point in time and space. Since the 1970s, however, it has been obvious that the expansion of the Universe did not follow the pattern that continuous expansion required. The only reasonable alternative was to doubt the cosmological red shift and to suggest that objects at high red shift are much less distant than calculated from a cosmological red shift.

The evidence that the red shift is cosmological, however, accumulated rapidly. For relatively nearby galaxies for which the distance can be measured directly by observing the brightest stars in the galaxies, or by observing Cepheids, or by observing globular clusters, it was evident that the red shift did indicate the distance perfectly. For more distant objects, there was the simple observation that the larger the red shift, the smaller and fainter the galaxy appeared to be. This is what scientists call empirical evidence—it is not definitive numbers, but it is strongly suggestive that galaxies with larger red shift were more distant. For many years the most distant objects observable in the Universe were quasars, and some scientists did challenge their red shifts as being at least partly noncosmological. Many quasars with large red shifts, however, had large numbers of absorption lines in their spectra. This is consistent with being seen through more nearby galaxies (in some cases, dozens of systems of absorption lines are seen at different red shifts), which again suggests strongly that quasars are at the distance their red shifts indicate. Similarly, more and more quasars were observed to be embedded in galaxies and surrounded by a cluster of galaxies at the same red shift as the quasar. Once again, seeing faint and obviously extremely distant galaxies at the same red shift as a quasar served to verify that the red shift of the quasar was correct.

For most scientists, the most conclusive proof has been from supernovae. Supernovae of type Ia have light curves and luminosities that are almost identical. Type Ia supernovae have been observed out to a red shift of 1.7, out to about 85 percent of the distance to the edge of the Universe. Put another way, such a supernova exploded when the Universe was just 15 percent of its present age. At this velocity of recession, time dilation becomes an important effect. We find that the light curves of distant supernovae are genuinely stretched out and happen in slow motion, exactly as we would expect if their enormous velocity makes time slow down.

Thus, all the effects that we see with increasing red shift are consistent with an expanding Universe that started at a definite point in time.

The Formation of the Elements

A great deal of time and effort has been dedicated to studying how the elements that make up the Universe formed. Of what are often, but incorrectly, called the 92 naturally occurring elements,[8] only 2 were formed in the Big Bang; all the other elements have been formed later in stars. All element formation in the early Universe took place during the first three minutes. The theory of the Big Bang states that initially the temperature of the Universe was so high that even protons and neutrons could not survive; all the mass of the Universe was in the form of quarks, and energy was in the form of high-energy photons. It was not until the Universe had cooled to approximately 3 trillion degrees that the quarks were able to come together as protons, neutrons, and other composite particles that together are known as "hadrons." The result was something not dissimilar to what is sometimes called a rich Greek alphabet soup.[9] The temperature and density of this gas were so high that no elements could form, or at least the only element that existed as such was hydrogen, due to the fact that there were many loose protons and a hydrogen nucleus simply consists of a proton.

The elements could only form during a brief space of time from one to three minutes after the Big Bang when the temperature had dropped sufficiently to permit elementary particles to join together, but the temperature and density were still high enough to permit the nuclear reactions that go on in stars. Over this two-minute interval, first protons and neutrons combined to form deuterium, and then, by process of adding protons, first tritium and then helium were formed. After three minutes had passed, the temperature and density of the Universe had fallen so much that the formation of elements froze.[10] Thus, until the first stars formed, the proportion of hydrogen and helium in the Universe was set at the exact amount that existed three minutes after the Big Bang, and even 13 billion years of star formation has changed it by only a small amount. This is 23 percent of helium by mass, or approximately 1 helium atom for every 10 hydrogen atoms. This quantity equates with a high degree of accuracy to the observed amount of helium in the Universe.

The CMB

For many scientists the discovery of the CMB marked the beginning of the end for rival cosmologies to the Big Bang. Although alternative suggestions have been made that interpret it as scattered light from distant dust, these have been unconvincing. The reason is the amazing degree of evenness of the temperature of the CMB, which only varies from point to point on the sky by a few millionths of a degree.[11] This evenness is consistent with it being produced from a highly homogenous gas such as the Big Bang would have been at the moment that it became large enough and tenuous enough to be transparent, allowing radiation from the early Universe to be detected.

The discovery by the Tenerife Experiment at Teide Observatory (Tenerife, Spain) and almost simultaneously by the COBE satellite of faint structures in the CMB that are the first large structures in the early Universe after the Big Bang, which would later condense into the first clusters of galaxies, has allowed astronomers to start to close the gap in our knowledge of the time between the most distant known galaxies and the Big Bang. The discovery of these structures has allowed the initial formation of galaxies to be studied.

We can see a scheme of the history of the Universe that the Big Bang and the CMB give us in figures 9.2 and 9.3. After the Big Bang, there was a brief period of what is termed *inflation*, in which the fabric of the Universe expanded with enormous rapidity at a velocity that was apparently far greater than that of the speed of light, although this velocity has no physical significance.[12] The CMB marks the moment 380,000 years after the Big Bang when electrons and protons could finally unite to form atoms and allow the Universe to be transparent. The CMB is, effectively, the boundary of the impenetrable fog that existed in the Universe beforehand. By this time, the first inhomogeneities were forming in the Universe that would later turn into clusters of galaxies. We now know that the first stars formed unexpectedly rapidly after the Big Bang. About 150 million years after the Big Bang, it seems that the first protogalaxies and the first stars were forming (figure 9.3). Over the next 13.6 billion years, these first irregular protogalaxies evolved into the great star systems that are the galaxies of the modern Universe.

The Finite Age of the Universe

If the Universe started with a Big Bang, it must have a finite age. In other words, it must have started at a certain point of time in the past. There are many ways of trying to calculate this age, such as looking at the ages of the oldest stars, or extrapolating backward the expansion of the Universe. As we have seen in chapter 8, the different methods of estimating the age of the Universe now agree, to within their uncertainties, on the age of 13.7 billion years given by the measurements from the Wilkinson Microwave Anisotropy Probe, whereas supernovae give an age of 13.6 billion years. As we can tie down the exact moment when the Universe began with such certainty, and because different methods of estimating it agree with each other, we have strong evidence that the Universe has a finite age and that it began at a specific moment, as the Big Bang Theory predicts.

Why Not a Quasi–Steady State (Oscillating) Universe?

The quasi steady state is the modern name for the oscillating universe model. In it the Universe expands after a Big Bang, the expansion finally stops, and there is a big crunch when all the mass of the Universe returns to its point of origin. In this model we have a "steady state" because the universe always regenerates and reforms, but during each cycle the universe formed will change. For many years this was probably the most popular of the different options for the evolution of the universe, in part because it was more satisfying than an open universe model in that there would always be regeneration and rebirth.

For the oscillating universe model to be correct, the amount of matter in the Universe had to be large enough to counterbalance the expansion and initiate a contraction. Many cosmologists suspected that, in fact, the amount of mass would be exactly that necessary to stop the Universe expanding at infinite time ($q_0 = 0.5$), and detailed Big Bang scenarios were developed to explain why this was a logical result. It was obvious, though, that the amount of matter visible in the Universe was far inferior to the amount required to close it. However, there was also strong evidence that so-called dark matter—because it was invisible—did exist in considerable quantities. Clusters of galaxies were seen, from the movement of individual galaxies within them, to be far more massive than the number of visible stars suggested. Many galaxies were observed, from their velocity of rotation, to be more massive than could be accounted for by just stars.

Many suggestions were made about the identity of the dark matter. Some astronomers proposed that it might be in the form of black holes created early in the history of the Universe, or tenuous hot gas between the galaxies, or in the form of exotic particles, or hidden among the ghostly particles known as neutrinos. There are countless billions of almost undetectable neutrinos in the Universe, but they have previously been thought to be massless. If each neutrino had a mass, even if it was less than a ten-thousandth of the mass of the electron, there would be enough mass in all the neutrinos to close the Universe. There was great excitement then when observations of the burst of neutrinos from the supernova Sn1987a in the Large Magellanic Cloud suggested that the arrival of the neutrinos was spread out in time rather than simultaneous, as it would be if an instantaneous burst of neutrinos had been emitted from the supernova at the speed of light. This hinted that neutrinos had a small but measurable mass. Various studies suggested that it might be of the order of 10 electron volts ($10\,eV$) compared with the mass of an electron of half a million electron volts, or the proton with 938 million electron volts[13] (usually abbreviated to $0.5\,MeV$ and $938\,MeV$, respectively). Ten electron volts would have been just enough for the total mass of all the neutrinos in the Universe to close the Universe. Later experiments cast grave doubt on these conclusions. Recent results in particle physics have suggested that there is a measurable neutrino mass, but that this is in the range from 0.1 to $0.5\,eV$, far below the amount of dark matter required to close the Universe.

The amount of dark matter in the Universe is far below what is required to stop its expansion. Summing all the dark matter and visible matter, we still only get about 30 percent of the critical mass. The best guess is that the mix is of about 1 gram of ordinary matter for each 6 grams of dark matter, whereas we would need there to be about 25 grams of dark matter for every 1 gram of ordinary matter to close the Universe and make an oscillating universe possible.

Cosmological Enigmas

How Certain Are We about the Big Bang?

Fifty years ago the Big Bang seemed far from certain, and a real alternative model existed. Today, however, the theory has dispatched all apparent challenges, and there is barely a cloud on the horizon for it. There is such solid evidence of the expansion of the Universe and the cosmological red shift that it is almost impossible to doubt it. The biggest challenges are in understanding the first moments of the Big Bang and, in particular, why it was so homogeneous and how the structures that we see in the modern-day Universe came to appear. Deep maps of the distribution of galaxies in the Universe show that they have a frothy structure with voids and relatively dense walls; at present, the biggest challenge that the Big Bang has is possibly to explain how this frothy structure came about when maps of the CMB show the expansion to have been so even in its first few hundred thousand years. If the Big Bang theory were to be overturned, it would require a discovery or discoveries of stunning proportions, and at present such a thing does not look likely.

SUGGESTIONS FOR FURTHER READING

Popular Books

Steven Weinberg, *The First Three Minutes: A Modern View of the Origin of the Universe* (New York: Basic Books, 1994).

> *The successor to the classic book* The First Three Minutes, *first published in 1977, this book was written by the Nobel laureate Steven Weinberg to explain the early history of the Big Bang until the end of element formation. Although recent discoveries have overtaken some of the details in the book, not only recent discoveries about the expansion of the Universe, but also discoveries in particle physics itself such as quark physics, this is still the best and most lucid account that exists of the Big Bang itself. It is not the easiest popular book to read, but is a good starting point for understanding the Big Bang.*

Michael D. Lemonick, *The Light at the Edge of the Universe: Dispatches from the Front Lines of Cosmology* (Princeton, N.J.: Princeton University Press, 1993).

> *This book discusses cosmology and theories of the Big Bang. The author shares many of the anecdotes that have accrued from many years of research into cosmology and a deep personal knowledge of the personalities involved, taking the reader behind the scenes in research into cosmology. Once again, recent results*

have overtaken some of the issues discussed in the book, but it is still a first-class insight into how astrophysicists think and work.

More advanced reading

H. Arp, "A Connection between the Spiral Galaxy NGC 4319 and the Quasi-Stellar Object Markarian 205," *Astrophysical Letters* 9 (1971): 1.

> *Halton Arp draws attention to the pairing of the galaxy NGC 4319 and the quasar Markarian 205 and the possible luminous bridge connecting them. It is an interesting read, whether or not you find the arguments presented convincing, and is valuable as a historical paper for the images contained and the discussion of their processing and enhancement in the days before CCDs and digital enhancement by computer.*

John N. Bahcall, Buell T. Jannuzi, Donald P. Schneider, George F. Hartig, and Richard F. Green, "The Near-Ultraviolet Spectrum of Markarian 205," *Astrophysical Journal* 398 (1992): 495–500.

> *Very definitely not an easy read, this study made with the Hubble Space Telescope demonstrated that Markarian 205 was observed through the galaxy NGC 4319 and thus behind it. For the overwhelming majority of scientists, this provided final and conclusive proof that the red shift of Markarian 205 does genuinely indicate its true distance.*
>
> *Available in the Internet at http://articles.adsabs.harvard.edu//full/1992ApJ . . . 398..495B.*

On the Internet

Wikipedia: The Big Bang
http://en.wikipedia.org/wiki/Big_Bang

> *Despite having its very public critics, the Wikipedia has gained an increasing reputation for its quality and completeness. The article on the Big Bang is thorough and detailed, and includes the very latest discoveries. The text itself is accessible to the general reader, although it uses terms that will not be familiar even to many astronomers, which may be followed up through a large number of links in the text.*

Big Bang Theory
www.big-bang-theory.com/

> *This short but interesting page is published anonymously. It gives a brief overview of the Big Bang theory and some of the great philosophical questions that surround it, including some of the misconceptions*

Cosmological Enigmas

that even many astronomers have about the Big Bang, what there was before the Big Bang, and the relationship between the Big Bang and the possible existence of a universal creator.

Errors in the tired-light theory
www.astro.ucla.edu/~wright/tiredlit.htm

This page is dedicated to refuting one of the alternative cosmological models, the idea that red shifts are not Doppler shifts, but rather due to light getting tired as it travels through the cosmos. The page is quite mathematical, but gives a series of observational proofs of the Doppler interpretation of the red shift and detailed criticism of alternative cosmologies.

NGC 4319 and Markarian 205
http://heritage.stsci.edu/2002/23/supplemental.html

A description of the pairing of galaxy and quasar that has been such a famous test case for the skeptics of the red shift. The page is presented by Roger Knacke of Penn State Erie and contains a series of nice medium-resolution images of the pairing, both earthbound photographs and images from the Hubble Space Telescope, along with annotations and interpretation of what they show.

Halton Arp
http://electric-cosmos.org/arp.htm

This page is dedicated to supporting the theories of Halton Arp regarding his doubts about the reality of the cosmic red shift. The page deals in some detail with various key examples of alignments of galaxies and quasars where the two objects have completely different red shift. Decide for yourself whether you find the arguments presented convincing in the light of the evidence presented in this chapter.

CHAPTER 10

What Is There Outside the Universe?

U p to now we have concentrated on questions with factual, if
sometimes controversial answers. To finish our journey to the
limits of knowledge and the cosmos, we need to reach into the
realm of the philosophical, well beyond the point at which science ceases
to have definitive answers.

The "universe" assumes a single cosmic all, what we might term "all
of creation." The idea that there might be multiple, parallel universes has
been a favorite theme of science fiction for many years. In some places
the term *multiverse* is used to describe the existence of multiple universes.
We can imagine the universe in which we live as a soap bubble; we are
inside the bubble. Is ours the only bubble that exists, or is it just one tiny
part of a huge mass of foam? If there are many bubbles, can we communicate with them? Can we even prove that they exist? Is there any limit
to the possible number of universes that exist?

To look at these questions let us start by looking at some of the people who have imagined this possibility.

Imagining a Multiverse

For many years the subject of multiple universes was the exclusive province of science fiction. Science fiction is an interesting medium, in that it allows great scope to writers and allows them let their imagination fly, sometimes imaging how things might be, but more often thinking "what if . . . ?" and starting from there ("what if time travel were possible," "what if travel to the planets were easy," etc.). It is a popular misconception that science fiction writers try to predict the future; in fact, the "what if . . . " variety is far more prevalent. Good science fiction writers attempt to make their science plausible—an extrapolation of the scientific reality of the day—whereas writers who do not do this deal more in fantasy. That is, the latter avoid the harsh reality of the physical laws that forbid traveling faster than light or the fact that a human turbo laser gunner cannot conceivably aim successfully at attacking fighters who travel at a speed of several miles per second on his or her own, without computers and radar to control the shots (even in World War II gunners rarely managed to hit aircraft traveling at only a few hundred miles per hour until radars were slaved to the guns); they imagine things that cannot be in our Universe. People tend to confuse fantasy and science fiction, but there is a huge difference between a space opera such as *Star Trek* or *Star Wars*, where science is liberally mixed with fantasy,[1] and pure science fiction in which authors try to limit themselves to the physical laws of the Universe and to things that could conceivably be one day or *could have been* if history had been slightly different.

Science fiction writers are a singular breed in that they are often highly qualified people who may well be professional scientists. Although many people turn their noses up at it as being some kind of prostitution of writing, good science fiction is often packed with good science or, at the very least, sensible extrapolation of science. One of the earliest pioneers of science fiction and the inventor of the "space opera" style with round-the-Galaxy action was Edward Elmer Smith. Known to his readers as E. E. "Doc" Smith, he received a Ph.D. in chemical engineering from George Washington University in 1919 and wrote some of the pioneering space operas that are still regarded as classics. These include the Skylark series, which he started in 1928 and which was the first serious writing on interstellar travel, and the still extremely popular Lensman series, the first book of which was published in 1948.[2]

Isaac Asimov was a biochemist who had a university position for much of his writing career and still lectured occasionally in the faculty even after he decided to dedicate himself exclusively to writing. Arthur C. Clarke was an engineer who obtained a first class degree at Kings College (London) and who was deeply involved in the development of the first blind-landing radar. Harry Turtledove, author of the alternative history science fiction World War series has a Ph.D. in Byzantine history from UCLA, having first started a degree at Caltech. Michael Crichton, writer of several famous science fiction novels, qualified as a doctor at Harvard Medical School. Jerry Pournelle (as in "Larry Niven and Jerry Pournelle," authors of superb science fiction novels such as *Footfall* and *Lucifer's Hammer*) is a remarkable polymath who has an astonishing four advanced degrees in psychology, statistics, engineering, and political science, including no fewer than two Ph.D.'s. He worked in both the Mercury and Gemini programs in the 1960s. Vernor Vinge, author of the Bobbles series of novels, *The Peace War* and *Marooned in Real Time*, was an associate professor in the Department of Mathematics at the University of California at San Diego. And, of course, Carl Sagan, the author of *Contact*, was a full professor at Cornell University until his untimely death.

The concept that there may be multiple universes in which all possible variations of history are being played out has given enormous scope for writers. Characters who swap universes accidentally, or by design, became a standard plot device. In some cases the idea is used in a comic way, allowing the protagonist of the story to get into ridiculous situations. A classic example is *What Mad Universe?* by Fredric Brown, in which the unexpected return to Earth of a failed first attempt to hit the Moon with a rocket probe blasts the hero of the story into a crazy parallel universe.[3] Another example is *The Stainless Steel Rat Wants You* by Harry Harrison, in which Harry Harrison's comic galactic policeman attempts to transport an invading alien army into a parallel universe where they will be someone else's problem.

The concept of a multiverse has also formed an integral part of both popular science fiction such as *Star Trek* and more serious science fiction too. In the *Star Trek: The Next Generation* novel *Q Squared*, Captain Picard must navigate his way through a series of different alternative universes to save the *Enterprise* and its own universe. Better known to the millions of Trekies around the world, though, is the episode "Mirror, Mirror" in which Captain Kirk and various of his crew end up being transported accidentally into a parallel universe in which their mirror-image selves have totally different and savage behavior.[4]

The classic example of a great science fiction writer using the concept of par-

Cosmological Enigmas

allel universes is undoubtedly Isaac Asimov in his novel *The Gods Themselves.*[5] Asimov imagines a situation in which a dying race in another universe contacts the Earth to exchange electrons for positrons between the two universes and thus generate an apparently infinite supply of free energy for the two universes. The poor scientists on Earth do not realize that the exchange will eventually destroy our planet as the laws of the alien universe are gradually diffusing into our solar system along with the "free" positrons and, when the new laws reach the Sun, they will accelerate nuclear fusion until the Sun goes supernova (moral: when you are getting something apparently for free, that looks too good to be true, then it *is* too good to be true).

The idea of multiple universes is not at all new. The concept played an important part in the writings of E. E. "Doc" Smith in the 1940s and early 1950s, even before the Big Bang theory was fully established scientifically, with the hero of the Lensman series, Kim Kinnison,[6] sent on various occasions (usually involuntarily) to an alternative universe in which the normal laws of physics that we know do not apply.

Could a multiverse as science fiction writers have described it have a genuine physical meaning? Or is it just the invention of a highly active imagination with no place in the real universe?

The Limits of Our Universe

Before we can look outside our Universe, we need to look first at its limits. This will help us to understand some of the problems involved in the idea of a multiverse. Despite the fact that modern discoveries suggest that our Universe will have an infinite and limitless future—in other words, it will continue expanding forever—it has a definite limit in the past. A consequence of the Big Bang is that the current limit of the Universe is 13.6 billion light years away. What happens when we try to look 13.6 billion light years into the distance?

The most distant known galaxy is at a red shift of 6.3. This would put it at 95.2 percent of the distance to the edge of the Universe, or at a distance of 12.9 billion light years. This marks the limit of what can be observed with ground-based telescopes or the Hubble Space Telescope (HST). With a bigger telescope like the European Southern Observatory's planned 100-meter-diameter Overwhelmingly Large Telescope (OWL) or the Sweden-Spain 50-meter-diameter

Euro-50 Telescope, or the James Webb Space Telescope (JWST), which will replace the HST at some time in the next 10 years, we will be able to observe more distant and fainter galaxies still. This is one of the most fundamental scientific reasons for requiring a telescope such as the Euro-50, which will be five times larger than the largest telescope currently in existence; without it, it will be difficult or impossible to satisfy our curiosity about the first billion years of the existence of our Universe and study the very first stars and galaxies to form.

As our knowledge of the most distant—that is, the youngest—parts of the cosmos has increased, it has become obvious that we need to understand the first billion years of the Universe far better to be able to understand how it turned into the Universe that we know. In particular, we know little about the formation of those first stars and galaxies. As these earliest galaxies evolved into the galaxies that we know, it is of great importance to understand their formation and early evolution. Similarly, observing extremely distant supernovae will give us fundamental information about the first stars and also about the expansion of the Universe in its early history.

For astronomers, telescopes like the Euro-50 and the JWST will be cosmology machines that will reveal the early history of the Universe and explain how it came to be the way it is. These telescopes should be able to detect the very first galaxies to form some 300 to 400 million years after the Big Bang and will also be able to study in much greater detail galaxies that are a little closer and still in the process of evolving into the present form. As important will be the ability to study supernovae in galaxies that are only a few hundred million years old. To date, the expansion of the Universe has been studied only out to a red shift a little more than one using supernovae; to understand how inflation has affected the evolution of the Universe, we need to be able to observe much fainter and more distant supernovae.

If a giant telescope can take us back to perhaps 300 million years after the Big Bang, if we make a larger telescope still, perhaps on the far side of the Moon, will it allow us to see back to the Big Bang itself and before? The answer is no. However large the telescope, we will be able to see the first stars forming about 200 million years after the Big Bang, but before that, nothing. The reason is that the period between 380,000 years and 200 million years after the Big Bang is termed the *dark ages*—there were no stars, there were no galaxies to see. So, if we cannot see stars and galaxies, why can we not see the Big Bang itself? It would have been intensely brilliant, so it should be clearly visible.

Cosmological Enigmas

The problem is twofold. First, we have the problem of the "photosphere." Like the Sun, the Big Bang was a huge ball of hot gas. We would expect the Sun, being gas, to be transparent, but clearly it is not. The reason is that, at a certain density and temperature of the gas, it becomes opaque—so opaque that it takes a photon generated in the Sun's core a million years to escape into space. The boundary where the Sun becomes opaque is called the photosphere and marks the limit of the visible disk—we cannot see deeper inside the Sun than this photosphere. Like the Sun, the Big Bang had a photosphere. For our Universe this photosphere comes 380,000 years after the Big Bang. Before that, the Universe is completely impenetrable in the same way that the Sun's core is invisible and can only be studied indirectly. Cosmologists call this photosphere *first light* because it is the first moment in time when light could escape from the expanding fireball and be detected.

So, even if we cannot see earlier than 380,000 years after the Big Bang, we should still see its brilliant glow from a moment in time when it was still at a temperature of thousands of degrees. So why do we not see the sky glowing white hot from the Big Bang, if only dimly in the far distance, 13.6 billion light years away? Here the second problem kicks in; we *do* see the hot glow of the Big Bang, but at a red shift so great that its light is shifted completely out of the visible part of the spectrum. In fact, so great is the red shift that the light is shifted right through to the radio range. In other words, it is the cosmic microwave background (CMB) that is so familiar to us. Theoretically, the CMB does emit some light in the visible part of the spectrum, but it is such an infinitesimal fraction of the energy emitted as microwaves that we have no hope of ever being able to detect it. In other words, it does not matter how big the telescope we use is; there is a point in space and time at some point between first light and the ignition of the first stars that even if there were something to observe—perhaps the first protogalaxies—it would be so hopelessly red-shifted that we could never detect it in the visible and can only hope to study it in the far infrared and with radio telescopes.

Thus, the physical laws of the Universe do set a fundamental limit as to how far into space we can see with optical telescopes. It does not matter how big they are, we can never reach the point of first light, or even approach it. Only with radio telescopes can we explore the almost featureless sky of the Big Bang's photosphere. What happened between then and the ignition of the first stars we can only guess at. Using physics and advanced mathematics, however, we can calculate what happened in the first few minutes after the Big Bang. So, does this give us any clues as to the possible existence of other universes?

Making the Leap

In recent years a small group of cosmologists has started to argue that the existence of a multiverse is a logical consequence of the Big Bang and that it is even supported by observational data. The pioneer of this school is a young Swedish-American cosmologist called Max Tegmark. Tegmark studied first in Sweden and then obtained a Ph.D. from the University of California at Berkeley.[7] He works now at the University of Pennsylvania with the data from the Sloan Digital Sky Survey (a survey of the sky using twin 2.5-meter telescopes at Apache Point Observatory in New Mexico) and also works with data for the CMB in an attempt to calculate the parameters that define the Universe.

Tegmark has defined four different possible levels of multiple universes—a multiverse—along with their physical justification (see figure 10.1). He suggests the question is not whether a multiverse exists, but rather how many levels of multiverse there are. It is fair to say, though, that not all his colleagues would agree with such a strong declaration.

Level 1 Multiverse

Level 1 is a consequence of the model of inflation. It supposes that when the expansion of the universe started, many individual volumes of space were enclosed. Each of these volumes is what we can call a "Hubble volume"—equivalent to our visible Universe. In this vision of the multiverse, space is like a huge mass of soap bubbles, with each bubble enclosing an individual universe. Within any one universe, one single Hubble volume, the furthest that you can see is to the boundary 13.6 billion light years away, given that the Universe was totally opaque closer to the Big Bang. Beyond that, there is the Big Bang itself and, beyond that, further Hubble volumes. However, the furthest objects from us within our own Universe are not actually at 13.6 billion light years distance because the Universe has expanded a great deal in the time since the light left them; in fact, the most distant objects are now about 40 billion light years away.

In this multiverse, different Hubble volumes contain copies of our own Universe. I am sitting at a desk in my study, tapping away on my laptop, while listening over the Internet to a BBC commentary on the current England versus Australia cricket match. This has been reproduced in another Universe exactly as now, but, instead of pausing a moment and lifting my hand to my chin as I ponder how to continue, I have, in the alternative universe, perhaps decided to turn and look aimlessly out of the window instead. You, the reader, also have a copy in this alternative reality who may have decided to get a cool beer rather than a coffee before continuing to read, or to cross rather than uncross your legs. How far away is the nearest exact copy of ourselves in another universe? The answer is spectacular: Tegmark estimates that they are installed $10^{10^{29}}$ meters away (this is a 1 followed by 10^{29} zeros), although if planet formation and evolution are as inevitable as some experts feel, there may be an identical copy of yourself far closer than this because the dice are "loaded" in favor of producing planets and life and thus in beating the odds. For those who like to think on a grander scale, at the modest distance of $10^{10^{91}}$ meters away, there will be an entire volume of space 100 light years across identical to that around our Sun, whereas some $10^{10^{115}}$ meters away we should find an entire Hubble volume identical in all respects to our own Universe at this present moment in time.

This is the simplest type of multiverse and is due to the simplest type of Big Bang. In it, physics is the same in all Hubble volumes; the only difference is that what physicists call the initial conditions are different (in other words, like the butterfly effect, where a butterfly flapping its wings in China may set in course a series of events that gives rise to a storm in New England, an apparently trivial difference in a universe compared to our own gives rise to a substantially different evolution of events later).

A multiverse of this type assumes that space is infinite, although we cannot see an infinite distance into space. In it, there are an infinite number of individual bubbles that have come from the initial inflation of the universe. This also gives rise to that phenomenon so beloved of science fiction writers in which every possible variant of every possible event, however unlikely, is taking place somewhere in space.[8]

There is strong evidence from observational cosmology that space is infinite and thus that the multiverse model is correct. Observations of the CMB show that the distribution of matter in the Universe is extremely even, something that was obvious from ground-based data and was later confirmed by satellites such as COBE and the Wilkinson Microwave Anisotropy Probe. This means that the Big

Bang itself was an extremely smooth event with almost perfect symmetry. A second proof of this is shown by the Sloan Digital Sky Survey (SDSS). The SDSS has mapped the distribution of galaxies around the Universe. We know that on *small* scales there is structure in the Universe: galaxies, clusters of galaxies, and even superclusters of galaxies. However, the SDSS has shown that, when we go to much larger scales, the distribution of matter in the Universe is extremely smooth with no important large-scale structure detectable. There are no coherent structures— that is, clustering—on scales larger than 0.1 percent of the radius of the Universe. In other words, even though galaxies form clusters and these clusters form together huge superclusters, which themselves may group into identifiable supersuperclusters, there comes a point at which larger-scale groupings just do not happen and everything blurs together into a single amorphous mass. When we go to 1 percent of the radius of the Universe, the level of variation in the amount of matter from point to point is 1 percent or less (in other words, a certain volume of space in one point of the Universe may have 100 clusters of galaxies, whereas the identical volume in another part of the Universe will have 99, or 101—not exactly a big variation from point to point).

This evenness of the Universe argues strongly that space must be infinite. Even though our Universe, the individual Hubble volume that we are inside, is not infinite, there must be an infinite number of other Hubble volumes outside ours to give this evenness in the distribution of matter on large scales with our own Universe. So, even though we cannot observe the universes beyond our own—they are hidden beyond the boundary imposed by first light in our Universe—they make their presence known by affecting the distribution of matter in our own Universe.

If the Universe was so even in its initial stages, how did it turn into the mass of complicated small-scale structure that we see today and how did different universes come about?

The smoothness of the Universe has been one of the biggest problems facing cosmologists over the past three decades. It is hard to understand how the Universe got from such an incredibly even and symmetrical Big Bang to such a complex structure—at least on small scales—in the modern Universe in which stars, galaxies, and clusters of galaxies can exist. Cosmologists suggest that random quantum fluctuations in the expansion during the period of inflation after the Big Bang were responsible for everything we see now.[9] This mechanism produces tiny random fluctuations from point to point in space, which lead to the formation of an infinite number of individual universes, each with random initial conditions, which then evolve in their own way dependent on those conditions. This means that all possible universes develop, but ensures that most of them will be close to the most probable universe and presupposes that our Universe is similar to the majority of other universes in the multiverse.[10]

Level 2 Multiverse

Level 2 supposes that, after the Big Bang and the brief period of initial inflation, different regions of space with slightly different physical laws were enclosed in separate volumes of space. Or, to put it more accurately, the equations of physics are the same, but the constants that go in them are different.

What does this mean in practical terms? Our Earth has a certain mass and diameter, and that leads to a certain strength of pull of gravity: an escape velocity of 11.1 kilometers per second. Suppose we go to a different region of the multiverse; in this region the gravitational constant may be double, or 10 times as large so, instead of an escape velocity of 11.1 kilometers per second, our own Earth might have an escape velocity of 22.2 kilometers per second, or even 111 kilometers per second. Or, to take another example, the strong and weak nuclear forces might be different in another part of the multiverse. This would have some fundamental effects. If, for example, the weak nuclear force, which controls beta decay (see figure 10.2) were stronger, radioactive elements that decay by converting a neutron into a proton and emitting a positron would decay far more rapidly and would be more radioactive.

Similarly, in a different part of the multiverse, the strong nuclear force, which binds atomic nuclei together, may be very much stronger than it is in our Universe—the same force, the same physical laws, but simply a different strength. This effect led to the detection of the parallel universe in the book *The Gods Themselves*, when a phial of stable tungsten-186 on a physicist's desk suddenly turns into a highly radioactive and unstable isotope of plutonium-186. The aliens from the parallel universe had exchanged the tungsten for an isotope of plutonium that was stable in the parallel universe but became extremely unstable as the laws of our Universe started to affect it. This exchange of material initially seems to be totally favorable to both sides, with the Earth receiving large quantities of seemingly free energy and the aliens receiving an energy source to replace their dying sun. But, alas, all is not as it seems; the strong nuclear force is also different, allowing stars in the parallel universe to generate copious energy from hydrogen fusion, even though they are no bigger than Jupiter. The consequence is that, as the different strong nuclear force from the received material infiltrates our Universe and spreads out, when it reaches the Sun it will cause the solar fusion reactions to accelerate out of control and the Sun to explode as a supernova (giving the parallel universe an even better supply of free energy, although with somewhat unfortunate consequences for the unsuspecting planet Earth).[11] Asimov's hero finally saves Earth by surveying other universes in the multiverse until one is found that counterbalances the effects of the first and restores our physical laws.

These examples show how the same physical laws can give completely different results in different universes of a multiverse just by changing the physical constants that control their strength. In our universe, the set of laws and physical constants that we have oblige stars to be accumulations of hydrogen that are a minimum of 50 to 100 times the mass of Jupiter even to be a dim red dwarf, while a star like the Sun is 1,000 times the mass of Jupiter. In an alternative universe we might find that an object the size of Jupiter would be a massive superstar,[12] or alternatively that something the size of our Sun would be a cold, dark planet according to how strong or weak the strong nuclear force is and how much mass is thus required to initiate fusion reactions. Similarly, in a universe where the constant of gravitation is much stronger, we might find that our own Universe would be massive enough to be strongly closed and collapse back in on itself. Seemingly trivial changes in the rules can thus lead to universes that are totally different from each other, even if the laws of physics that govern those universes are constant.

A level 2 multiverse will result if the initial inflation of the universe is chaotic and multiple regions of the Big Bang are enclosed in individual bubbles. The laws of physics are fixed as they freeze out during the initial expansion of the Big Bang. If the inflation is chaotic, many bubbles of space can form, each freezing out at a slightly different moment and with slightly different versions of the laws of physics, manifested as different "universal" constants of physics to the ones in our own Universe. This idea makes a level 2 multiverse attractive as one can imagine intuitively how it might come about easily in the Big Bang. Is there, though, any reason to imagine why it might have come to happen in reality?

Although the Big Bang theory has been extremely successful in explaining the early history of the Universe, there are a number of questions that remain to be

Cosmological Enigmas

answered. One of the key questions is the presence, or rather the apparent absence, of a particular particle that theories state should have been created in huge quantities in the Big Bang.

Particle physicists have been searching for many years for evidence of a particle known as a magnetic monopole. The existence of this particle was first predicted by Paul Dirac in 1931. In nature we know that any magnet we can make, however tiny, must have both a north and a south pole. Dirac suggested that there must be particles in the universe that are isolated north or south poles. Theory predicts that untold billions of extremely massive magnetic monopoles should have formed in the Big Bang. Despite many years of searches, physicists have not found solid evidence of the existence of monopoles, each of which would have a mass of approximately 10^{16}GeV.[13] The fact that there are no obvious monopoles in our Universe is a problem for Big Bang models and suggests that maybe they are simply elsewhere than in our Universe, or so spread out as to be undetectable.

Another problem is the flatness of space itself. Observations show that the Universe is extremely flat on a large scale, whereas the models suggest that it should be strongly curved. What this means is that instead of having a mass close to or superior to the critical mass that would close the Universe (the most logical models suggest that the mass should be almost exactly the critical mass) and thus space that would be strongly curved by gravity, the mass of the Universe is far below the critical mass and thus space is unexpectedly and quite unreasonably flat. Similarly, the CMB is also quite unreasonably flat: how can completely different parts of the Big Bang that were not in contact be at exactly the same temperature? All these problems can be solved by inflation. Inflation also provides a logical way of generating a multiverse. In the popular model known as chaotic inflation, as the inflation continues in some regions of space, it will cease in others, closing off bubbles of space, while other regions inflate even more rapidly due to quantum fluctuations. This produces a chain reaction in which each bubble of inflation leads to the generation of further bubbles. In essence, we have a model in which chaotic inflation leads to an infinite number of enclosed bubbles of space, each of an infinite size, in a process that continues for an infinite amount of time (in other words, the multiverse has no beginning and no end).

Why believe in such a multiverse? Supporters turn to what they call "fine-tuning." They argue that our Universe is inherently improbable. For example, only 1 in 1,000 universes should have a CMB as smooth as in our Universe. However, there are many more examples in physics of how, were the laws of physics written just slightly differently, our Universe would be impossible. A 4 percent weakening of the electromagnetic force would allow two protons to combine directly into a helium nucleus, for example, which would short-circuit the slow intermediate steps that slow down nuclear reactions in the Sun and other stars. The result would be that the Sun would become immediately and violently unstable. Even the smallest red dwarf would blaze as a supernova. Make the weak nuclear force significantly weaker, and reactions in the Big Bang would proceed so rapidly that all the hydrogen would be converted almost instantly into helium. Make protons just 0.2 percent more massive, and they would become unstable and decay into neutrons, so there would be no stable atoms in the Universe. In other words, they argue that we only exist because many things are "just so" in our Universe. This fine-tuning of such fundamental aspects of physical laws to allow us to be possible argues that our Universe is just one of an infinite number of possible configurations that have been generated in a multiverse.

Opponents of the multiverse concept state that this is a misuse of the anthropic principle. Instead of "I think, therefore I am," they argue that we are saying, "I exist, therefore my universe is special." If life will only develop in a universe with certain physical laws, it is not a great surprise that we live in such a universe. In other words, opponents suggest we are obtaining false conclusions from skewed data that make a particular answer inevitable.[14] The argument over this point is heated and unlikely to arrive at a conclusion in the near future.

The battleground for this multiverse theory is thus whether we are skewing the data by picking just the peculiarities that make us possible to justify the fact that we exist!

Level 3 Multiverse

Level 3 of the multiverse is the one most endeared of science fiction and fantasy writers. It supposes that the random nature of quantum physics allows exactly the same physical laws to exist in all possible universes but that, at every moment in time when there is a branching of events, different universes peel off allowing all possible histories to occur. For example, you decide suddenly that you need something at the shop and leave just at the moment that the man or woman of your dreams is about to pass your front door. In one universe, you exist at exactly

Cosmological Enigmas

the right moment and coincide; in another, you are a second too early or too late to meet. In the universe where you meet, you ask the person to come and have a drink with you. In one universe the person says yes, and you move down one branch of history; in the other you are ignored or told to take a hike, and history moves down a different branch. At every particular branch point, all possible outcomes will happen in one universe or another.

The basis for this idea is that quantum physics does not allow absolute certainty. Everything is described as a wave function, which does not allow its position to be fixed absolutely. The most famous application of this is the Heisenberg uncertainty principle, which states that if you establish the position of a particle—say, an electron—with infinite precision, you will know its velocity with infinite uncertainty, for the product of the two errors cannot be smaller than a certain quantity. Einstein's famous rebuke to quantum theory was to state that "God does not play dice with the Universe," but despite his inability to believe quantum physics, we know that this was one of Einstein's few serious errors.

In this theory, the issue is whether physics is unitary. Every particle in the Universe has its behavior determined by a wave function described by the Schrödinger equation. This state evolves with time in a deterministic fashion. So far, so good. This situation, though, allows bizarre things to happen, such as a particle being in two places at the same time without violating the laws of physics.[15]

Great efforts have been made by physicists to check that different physical systems are unitary. So far, no exceptions have been found, and it is argued that even quantum gravity appears to be unitary. This gives a multiverse model that is mathematically simple. Physics is identical in every universe of the multiverse, so there are no problems with alternative universes in which the Sun is unstable, or we cannot form the elements required for life and spread them through space with supernova explosions. The only difference between the universes is the way that at every possible crossroads of history, however trivial, every possible course gets taken by history, each in its own universe. The result is no different to a level 1 or level 2 multiverse, so we are not complicating life, thus a level 3 multiverse could exist simultaneously with these other levels.

Level 4 Multiverse

The final type of multiverse that can be imagined is one in which physics itself is different in different universes. Why this universe and these physical laws instead of another set? Is there any particular physical reason bound up in the universe to state that the strength of gravity must fall off with distance as an inverse-squared law rather than some other law? Why not a universe in which the force of gravity gets *stronger* and not weaker as two objects separate?

In a sense, this is similar to the search for the ultimate question in *The Hitchhiker's Guide to the Galaxy*.[16] Physicists search for a Grand Unified Theory that will allow them to explain all of physics through a single law or equation. Fans of the *Hitchhiker's Guide to the Galaxy* will know that the answer to this particular question is 42, but nobody has the remotest idea what the question is that gives this particular answer; thus the answer, without exact knowledge of the question, with due deference to Heisenberg, is totally indeterminate and utterly useless.

Here we are no longer talking about subtle changes in physical laws that involve certain constants in the equations to be slightly larger or smaller; we are talking about completely different laws of physics that are totally different from the ones in our Universe. Essentially, if there is a single physical equation that determines how all the laws of the Universe work, why is there just that one law and no other possible ones? This was first asked by the physicist John Wheeler and later by Stephen Hawking, who pondered whether there should be a fundamental reason why the equations that govern our Universe should be as they are to the exclusion of all other possibilities.

The scientists who support this type of multiverse propose that our Universe is governed by a particular mathematical structure. That being so, there is no reason why we should not envisage other, different mathematical structures to describe the universe. That being so, if a particular mathematical structure exists that describes a possible universe, why should only the structure that leads to our universe be the only one that exists; why not all the other structures too? In other words, if a solution to a universe exists *mathematically*, that solution must logically and automatically exist *physically*, too.

Here the arguments become more complex and esoteric. Why assume that the Universe is a mathematical structure? The answer is that mathematics plays a weird part in physics and astronomy. In, for example, engineering or social sciences, there is always a degree of estimate in the results of any particular process; however hard you try, you can only get the numbers to a certain level of accuracy. In contrast, in astronomy, if you have good enough observational data, you can fix the numbers to the last decimal place that your observations permit without approximations at all. Mathematics just works amazingly well with physics. It has been argued dispassionately that there is no conceivable physical reason why mathematics and physics should dovetail so perfectly unless the Universe is a mathematical structure.[17]

If the Universe is a mathematical structure and if all possible solutions exist, then there must be an infinite number of universes, each with its own individual physical laws. This satisfies the physicists who worry about "fine-tuning," for it allows a universe to exist that is precisely tailored to our needs. All the convenient coincidences of the particular forces and constants of the Universe being conveniently "just so" are perfectly acceptable, while untold billions of other universes are not so lucky as we are with the numbers. It also solves the Wheeler-Hawking question because all possible physical laws will exist somewhere in the multiverse, and thus our laws are special only in being the ones that just happen to apply to our own little corner of the macrocosmic all.

Is This Science?

Critics of the multiverse theory argue that any theory that cannot be falsified (which means it cannot be demonstrated to be false, not that a scientist is cheating) is not real science. The kinder critics say that it is metaphysics rather than physics. Those less tolerant refuse to treat the subject as being science at any level at all and regard those who talk of multiple universes as engaging in the dissemination of science fiction (or, more callously, speaking of fantasy). If we cannot observe the alternative universes and cannot carry out an experiment that would prove their nonexistence, then to speak of them is not proper science.

A second line of argument refers to the fact that the simplest type of cosmos is one that supposes that there is just one universe; why complicate matters by adding additional unseen and unseeable universes? Adding an unnecessary complication to the universe is unscientific and violates the basic tenets of Ockham's razor, which states that, all things being equal, the simpler of two theories is most likely to be the correct one.

A further argument is that the laws of physics only allow one universe. This is a classic argument that was explored by Einstein. He wondered why the universe is as it is; why should it not have other laws, or even not exist at all? A grand unified theory of physics from which the laws of the Universe derive is still being sought by scientists like the Holy Grail. Such a grand unified theory could provide the missing proof. Will it give a single, unique solution? If so, precisely one universe—our own, with its observed physical laws—can exist. Thus, the critics state, once the unified theory exists and is shown to have only a single solution, any speculation about the possible existence of a multiverse will be totally meaningless. Conversely, if a grand unified theory were to give multiple solutions, that would provide strong evidence that the multiverse theory is correct.

Thus, for talk of a multiverse to gain widespread acceptance, its proponents must tackle head-on the problems of falsifiability and simplicity and demonstrate that cosmological models both demand a multiverse and can be tested properly to the satisfaction of the scientific community. Theoreticians like Max Tegmark have already commented that the smoothness of the CMB suggests that there is an infinite extension to space and thus that there is a multiverse rather than a universe, but, so far, this argument has not been strongly supported by cosmologists.

Can we ever prove (or, at least, *disprove*)[18] the existence of a multiverse? Certainly the possible existence of level 1 and level 2 multiverses can be refuted in the future. Much more sensitive and precise measurements of the CMB and of the large-scale distribution of matter in the Universe will provide stiff challenges for current models. We will see if the measurements of the curvature of space are consistent with a level 1 multiverse model. Similarly, these measurements will test the validity of the inflationary model of the Universe that leads to the possibility of a type 2 multiverse. Levels 3 and 4 are more challenging as they rely on larger leaps. Tegmark suggests that the success of the ongoing effort to build quantum computers would be a huge step toward supporting a level 3 quantum-generated multiverse. The level 4 multiverse seems the hardest to test at present, as this requires a unification of general relativity and quantum field theory, something that does not seem to be likely in the foreseeable future. Such unification would reveal whether there is just a single solution of the equations that govern the laws of the Universe or, on the contrary, multiple solutions are permissible, in which case the existence of multiple universes, each with its own physical laws, becomes more plausible.

Cosmological Enigmas

AT THE TIME OF THIS WRITING NO ONE CAN EITHER prove or disprove the possible existence of a multiverse or, as it is popularly known, parallel universes, although scientists can think of convincing reasons why they could exist within the laws of the universe as we understand them. This is one of those peculiar cases where science fiction has speculated with a seemingly crazy and impossible idea and, much later, science has started to take that idea seriously and demonstrate why it could possibly be true. Modern theories of the Big Bang, supported by the best available data, seem to tell us that multiverses could well exist in practice. Within 20 years much better data about the farthest reaches of our Universe may allow us to show that such theories are genuinely consistent with what we know about the Universe and, despite our best efforts, cannot be excluded. Even if we can establish that parallel universes are plausible, however, that is a far cry from being able to visit them by means of a malfunctioning transporter, or punching a hole in hyperspace. Travel to parallel universes will, almost certainly, remain the stuff of fantasy.

SUGGESTIONS FOR FURTHER READING

Science Fiction

Isaac Asimov, *The Gods Themselves* (Frogmore, St. Albans, Hertfordshire: Panther Books, 1972).

> *This science fiction story is far from being my favorite piece of Asimovana, but it describes brilliantly a level 2 multiverse some 30 years before Max Tegmark described it formally. The story is set in the mid-twenty-first century and is an excellent description of the differences between different universes in a level 2 multiverse. This is ideal as a gentle introduction to the idea of parallel universes within a multiverse.*

More Advanced Reading

Alan Guth, *The Inflationary Universe* (New York: Perseus Books, 1997).

> *Despite being a highly technical subject and not being an easy read, this is a fascinating and well-written book. Alan Guth explains theories of cosmology and the origin of the Universe and their relation to modern discoveries in elementary particle physics. The book shows how the theory of inflation came about and the predictions that it makes about the Universe and its structure. Chapters in the book cover such topics as the problems of the (missing) magnetic monopoles and grand unified theories.*

What Is There Outside the Universe? 191

Stephen Hawking, *A Brief History of Time* (New York: Touchstone Books, 1993).

> *This classic book unexpectedly became a great best seller. Stephen Hawking explores the theories of the universe from the Big Bang to grand unified theories and from black holes to quantum mechanics. In it Hawking explores the question of why the particular physical laws that govern our Universe should be as they are.*

Max Tegmark, "Parallel Universes," *Scientific American*, May 2003, 40–51.

> *This article describes the possible levels of a multiverse and the scientific evidence for them in the light of recent cosmological observations and discoveries. It is not a simple read, and the average layperson will need to take some parts of the description for granted, but it is a remarkably clear presentation of such a complex topic.*

> *A formal version of Tegmark's article was published as "Parallel Universes," in* Science and Ultimate Reality: From Quantum to Cosmos, Honoring John Wheeler's 90th Birthday, *edited by J. D. Barrow, P. C. W. Davies, and C. L. Harper (Cambridge: Cambridge University Press 2003) and can be found as a PDF file on the Internet at http://www.wintersteel.com/files/Shana Articles/multiverse.pdf.*

David Deutsch, "The Structure of the Multiverse" (2001).
http://xxx.lanl.gov/ftp/quant-ph/papers/0104/0104033.pdf

> *Definitely not an easy read, this text describes the structure of a multiverse based on the precepts of computational theory.*

On the Internet

Multiverse
http://en.wikipedia.org/wiki/Multiverse

> *Pieces in Wikipedia are not signed and one can never be sure who has made the contribution. In this case, though, the piece appears authoritative and is well referenced. It gives a fairly readable overview of the concept of a multiverse and its origins in science and philosophy.*

Notes

INTRODUCTION

1. The mirror is not circular, but rather approximately hexagonal, so the "largest diameter" means the largest span across the mirror from one side to the other. The *average* distance across it is 10 meters. The telescope's official first light ceremony was on July 17, 2007. The telescope now enters commissioning tests. The delay in this project has allowed the South African Large Telescope (SALT), a full 11 meters in diameter, to overtake it.

CHAPTER 1. How Are Stars Born and How Do They Die?

1. All three measurements used the technique of parallax. If you measure the position of a star in the sky with respect to the background of more distant stars, you will find that you see it in a slightly different position in, say, July to its apparent position in January. This is because in the six months between the two measures the Earth has moved halfway around the Sun in its orbit, and we are thus looking at the star from a slightly different direction. Knowing the diameter of the Earth's orbit and the angle by which the star's apparent position shifts—the parallax—it is a simple problem of trigonometry to calculate the distance to the star.

A practical example of how parallax works is to hold a finger in front of your face at arm's length. Close one eye and, with one eye closed, line the finger up with a book on a shelf, or a lamp post. Now switch eyes and see just how much the apparent position of your finger has changed. The further you hold the finger from your face, the smaller the apparent change in position will be when you change eyes.

2. Henderson actually carried out his measures in 1832–33, some five years before Bessell, but, due to his ill health, which led him to return to Scotland from the Cape Observatory, his results were only published afterward. In 1834, at the age of 36, he became the first ever Astronomer Royal for Scotland. However, dogged by ill health, Henderson died only a few years later, when still only 46.

3. At the time Isaac Newton was a student at Cambridge University, just finishing his degree. However, due to the last great epidemic of plague to ravage Britain, the university was closed, and Newton returned to his family home in Woolsthrope in Lincolnshire. For a prolific year he carried out many fundamental experiments free of the strictures of university. By 1669, with Newton at the age of just 26, his mentor at Cambridge, Professor Isaac Barrow, resigned the Lucasian Chair of Mathematics to allow his pupil to take it over. Apart from his experiments on light and gravity in 1668, Newton designed and constructed the world's first reflecting telescope.

4. This discovery was later to have a huge influence on my career. In 2006 I started to work for the European Space Agency's Herschel Space Observatory, a 3.5 meter infrared telescope named in honor of Sir William Herschel's discovery of infrared radiation.

5. The inventor of the Bunsen burner so well known in school chemistry labs.

6. Annie Cannon was born in Delaware in 1863 and joined Harvard Observatory at the age of 33. Wilhelmina Fleming was born in Scotland in 1857 and later emigrated to the United States, where she joined Harvard Observatory and was put in charge of the famous Henry Draper star catalog, a compilation of data on 225,300 stars. Apart from the basic information of the position and brightness of the stars, each of the stars had a spectral type on the new Harvard system. The work of classifying the stars fell mainly upon the shoulders of Annie Cannon, who carried out one of the most prodigious feats of hand cataloging in history, looking at the spectrum of every one of the stars and deciding what type it was. For many stars the spectral classification used continues to be that of Annie Cannon and the Henry Draper Catalog.

7. Why 32.6 light years? The reason is that this is exactly 10 parsecs, or the distance at which a star would show a parallax of exactly 0.1 arcseconds, making it a convenient round number. At a distance of 3.26 light years, a star would show a parallax of exactly 1 arcsecond, so this distance became known as a parsec as a contraction of "parallax of a second." Parsecs and light years are both regularly used by astronomers to measure distances. In cosmology the term Megaparsec is often used for a million parsecs to express the distances to galaxies.

8. James Jeans, *The Universe around Us*, 2nd ed. (London: Cambridge University Press, 1930). Perrine seems to have been a highly colorful character. He discovered nine comets and two of Jupiter's satellites, but made himself highly unpopular and, in 1936, retired after narrowly escaping assassination.

9. The error was a curious one. It was known that the Sun was losing mass at the rate of 4 million tonnes per second (you just use Einstein's famous equation $E = mc^2$—as you know how much energy the Sun is producing, because you can measure it, you can calculate how much mass

loss that energy output is equivalent to). Scientists thus divided the Sun's mass in tonnes by 4 million to calculate how many seconds that mass would last and took that as the age of the Sun! The fallacy in this is, of course, the fact that only a tiny fraction of the Sun's mass is converted into energy when its hydrogen is converted into helium (about 0.7%) and only a small fraction of all the hydrogen in the Sun is converted into helium anyway.

10. Bethe had been working as an acting assistant professor at the University of Tübingen in 1932–33, a job that he lost with the advent of the Nazi government in Germany. After emigrating to the United Kingdom in autumn 1933, he held positions at the universities of Manchester and Bristol before receiving an appointment at Cornell University. Only two years after joining the staff of Cornell, he was promoted to full professor. Like many exiled European atomic physicists, he ended up working at Los Alamos on the Manhattan Project. In 1999, at age 93, he gave a series of lectures on quantum physics, resuming the work that he had carried out over his 75-year career and demonstrated that even at such an advanced age he was still at the forefront of research in what is often regarded as a young man's field. These lectures may be found at http://bethe.cornell.edu/.

11. Carl Friedrich von Weiszäcker, a German physicist, had reached the same conclusions as Bethe almost simultaneously. Von Weiszäcker's paper reached the German-language journal *Zeitschrift für Physik* on July 11, 1938, whereas Bethe's paper was received by *Physical Review* on September 7. However, Bethe's paper extended an earlier published work on hydrogen fusion in stars that had been received by *Physical Review* on June 23; thus priority and credit is given, somewhat unfairly, to Bethe, while von Weiszäcker's contribution is largely forgotten by the scientific community.

12. The green color comes from the emission of light from oxygen atoms in the nebula. The dark-adapted human eye is sensitive to this color. Photographs, though, show a strongly red color. This is because the hydrogen in the gas cloud emits light principally in the red line of hydrogen alpha; a color emulsion is very sensitive to this color, but the human eye is not, particularly when dark adapted.

13. Less than 0.1 percent of the hydrogen in the star is in the form of deuterium, but in very low-mass stars there will be some energy generation from fusion of deuterium, even though there is not enough mass to start the p-p chain.

14. It was this figure that led to the estimate that stars like the Sun were a million million years old. Astronomers calculated how long it would take for the Sun to turn its 2,000,000,000,000,000,000,000,000 tonnes of mass into energy at a rate of 4 million tonnes per second and used that as an estimate of its age (see also note 8).

15. Chondrules appear to be material from the original nebula that condensed and solidified before being incorporated into the protosolar disk and accreted into the planets. Some of them survive today, unchanged, because they came to form part of the asteroids and have since come to Earth in meteorites. When we cut open a stony meteorite, we often find that it is full of these little balls of material with a different color and texture from the rest of the meteorite, like the currents inside a fruit bun. The chondrules may be from a few millimeters to perhaps a centimeter across. As such, they reveal the conditions in the cloud that gave rise to the formation of the Sun and the planets.

16. This aluminum-26 also played a huge role in the early history of the solar system because, as it is so unstable and radioactive, it was a source of enormous quantities of energy. Even asteroids down to a few tens of kilometers in diameter may have been melted inside by the decay of aluminum-26, giving them molten cores and separating the light and heavy elements in the same way that in the Earth the iron has sunk to the core and the lighter silicate rocks that form the Earth's crust have floated on top. This explains why some meteorites that fall to Earth are made of iron (they have come from the core of asteroids that have been destroyed in the past in collisions) and others are made from light, silicate rocks, which have come from the exterior, surrounding the core.

17. The group is known as the Seven Sisters because the ancients said that seven stars were visible to the naked eye. In modern times, though, only six stars can be seen easily with the naked eye. The seventh star, Merope, is known popularly as "the lost Pleiad" and may have faded over the last 2,000 years.

18. This somewhat odd situation comes from the fact that, when we look up at the night sky, we see an overwhelming majority of stars larger and much more luminous than our Sun. Of the 2,000 or so stars that a person with good eyesight will see on any dark night, only a handful are small and dimmer than the Sun, and not one is a red dwarf, despite the fact that the majority of all stars are red dwarfs. The reason for this is that a star smaller than the Sun can be seen only if it is relatively close by. If our Sun were moved to a distance of just 57 light years, it would be magnitude 6 and barely visible to the naked eye. Most red dwarfs, though, would have to be a lot less than 1 light year away to be visible to the naked eye. In contrast, the stars of Orion's Belt appear bright despite all being more than 1,000 light years away because even the least luminous of them is 18,000 times the luminosity of the Sun and could be seen with the naked eye to a distance of around 10,000 light years. We thus see preferentially the more luminous stars despite the fact that they are usually far more distant.

19. Astronomers over the years classed supernovae as being type I or type II according to the spectrum of the explosion. Later, type I was subdivided into types Ia, Ib, and Ic according to whether hydrogen or helium is seen (or not) in the spectrum. Only later was it realized that supernovae of type Ib and Ic are really identical to a type II supernova, but with the difference that the outer layers of the star have been lost before the final explosion, making the star appear to be a much hotter and bluer object than a red giant star. According to how much of the outer layers have been lost and how deep we see into the core of the star, it will appear as type Ib or Ic.

CHAPTER 2. How Do We Know That Black Holes Exist?

1. For the pedant, the exact value is 11.3 kilometers per second.

2. Technically, this is not quite true. Einstein established that no physical body can travel *at* the speed of light because at this speed its mass would become infinite. The special theory of relativity does not forbid velocities *greater* than that of light but, as we cannot conceive of any way of accelerating from slower than the velocity of light to faster than the velocity of light without *passing through* the velocity of light, the velocity of light is a giant and uncrossable wall. Physicists have

imagined particles called tachyons, which have the opposite problem: these are hypothetical particles with an imaginary mass (in other words, a mass that is a multiple of the square root of −1) that can only travel faster than light and that, as they slow, their (imaginary) mass becomes infinite as they reach the speed of light.

Science fiction writers have imagined all kinds of ways of tricking physics to exceed the speed of light, none of which are actually possible in reality. As I understand it, Captain Kirk and the Starship *Enterprise* manage this neat trick by warping space by some unknowable means, so that distances are massively reduced and thus the *Enterprise* travels at what is effectively many times the speed of light. A study by a Belgian mathematician published in 1999 even suggested how this might work in practice. You can find a report by the BBC on this at http://news.bbc.co.uk/1/hi/sci/tech/364496.stm.

3. Although the war on the western front, with its trenches, poison gas, tanks, and unbelievably bloody battles to gain just a few hundred meters of ruined ground, has become far more famous in history and even somewhat romanticized by those who had the good fortune not to experience it, the eastern front was, in its way, even crueler, having the bitter Central European winter to add to all the other hardships. That Karl Schwarzschild was capable of doing anything more than just try to survive in such an environment is astonishing; that he was capable of two huge breakthroughs in physics is quite unbelievable and demonstrates his remarkable qualities.

4. If you really need to know, they are a series of 16 coupled hyperbolic-elliptic nonlinear partial differential equations that describe the effects of any mass on the gravitational field. They are "nonlinear" because of the fact that all masses affect the very geometry of the space in which they dwell and thus curve space itself, although such effects are only seen when the mass is extremely large or its density is extremely high.

5. Just like the lines in an ordinary spectrum, each element emits x-rays of a certain energy. This allows astronomers to fingerprint rocks to identify the elements within them. This technique is now used widely to analyze the composition of the surface rocks of the Moon and planets, either at a distance from spacecraft or on the surface, like the x-ray spectrograph on Mars Pathfinder's Sojourner rover.

6. E. E. Salpeter, "Accretion of Interstellar Matter by Massive Objects," *Astrophysical Journal* 140 (1964): 796–800. This is what is often called a "priority" publication whereby a revolutionary new theory is commented in such a way as not to appear outrageous—and thus probably unpublishable—at the time, but which will establish the author's priority with the idea later when it becomes more respectable.

Edwin Salpeter is another of the generation of "lost" European astronomers from the 1930s who made his career in the United States. He fled with his parents from Hitler's Austria to Australia and studied physics at Sydney University. Later he obtained his Ph.D. in quantum electrodynamics with Rudolf Peierls at Birmingham University in the United Kingdom and finally went, like many exiles from Hitler's Europe, to Cornell, where he worked with Hans Bethe. He has worked in many fields of astrophysics and has also collaborated with his wife in neurobiology and with his daughter in epidemiology studies, as his remarkable autobiographic look at his career "A

Generalist Looks Back" explains. Unusually, it is published in one of the most hallowed of learned journals, *Annual Reviews of Astronomy and Astrophysics* 40 (2002): 1–25. This journal is dedicated to heavyweight articles from distinguished scientists who try to summarize knowledge in a particular field of astrophysics. A link to the article including his photograph and a summary of the text (but, regrettably access to the full text only for subscribers) can be found at http://arjournals .annualreviews.org/doi/abs/10.1146/annurev.astro.40.060401.093901?journalCode=astro.

7. This would be possible if the mass was at the lower end of the range allowing two neutron stars each just below three solar masses to sum five times the mass of the Sun between them.

8. Now, special telescopes on Earth are able to detect even photons that register as hundreds of TeV—terra electron volts where 1 TeV is 1 trillion electron volts.

9. After finishing her Ph.D. at Cambridge, Jocelyn Bell became Jocelyn Bell-Burnell and turned to x-ray astronomy. On an early appearance on the BBC's *Sky at Night* program I remember her reciting what she called the hymn of the x-ray astronomers: "Through the night of doubt and sorrow / Onward marches the pilgrim band / Counting photons very slowly / On the fingers of one hand." It takes a long time to count x-ray photons, and the higher the energy, the longer it takes. And, yes, Jocelyn Bell-Burnell is a most charming personality and a wonderful public speaker.

10. Both discoveries are reported in the IAU Circular at http://cfa-www.harvard.edu/iauc/ 04700/04782.html and 04783.html. There is no hint of what an important object the new x-ray nova would turn out to be.

11. Seyfert galaxies were first described by Carl [Keenan] Seyfert, at the Case Institute in Cleveland where he was teaching navigation to the armed forces. Apart from his teaching and work on secret military projects, Seyfert found time to carry out some astronomical research. He recognized that a class of galaxies with an unusually bright and stellar nucleus that he had been studying previously when at Mount Wilson Observatory also showed an unusual spectrum totally unlike normal galaxies. Seyfert, a remarkable polymath, died in an automobile accident in 1960, at the tragically young age of 49. After his death it was realized that Seyfert galaxies are intimately related to quasars; they were the first Active Galactic Nuclei to be recognized.

12. As Sir James Jeans wrote a few years later, astronomers initially found the idea of a massive, compact, white-hot star completely absurd, whatever the observational evidence for it, but a few years afterward, Albert Einstein suggested something that seemed even more doubtful, that is, the idea of the gravitational red shift, so astronomers decided to use one doubtful thing—the existence of white dwarf stars—to confirm successfully something that seemed even more doubtful—the prediction of the existence of a gravitational red shift. When Sirius B was found to show the gravitational red shift, both theories were confirmed in one fell swoop.

CHAPTER 3. Who Is the Strangest in the Cosmic Zoo?

1. The term *supernova* was first coined by the astronomer Fritz Zwicky in the 1930s.

2. It is often stated that one of the movers in this was the five-year-old son of two of the pioneer observers of quasars—Geoffrey and Margaret Burbidge—who on one occasion and with

the total innocence, lack of self-consciousness, and ability to hit the nail on the head that only a five-year-old possesses, asked his parents, "What's a crazy stellar object?"

3. For a start, less than 10 percent of quasars are actually radio sources, and a lot of them are not strongly blue.

4. The name OJ287 comes from: "O," the Ohio survey; "J," a radio source detected in the range of Right Ascension from 8 to 9 hours (0–1 hours = B, 1–2 hours = C, 2–3 hours = D, etc.); "2," the source is between Declination +20° and +30°; and "87," the decimals of hours of Right Ascension of the position (i.e., "87" = 0.87 = 0.87 × 60 = 53 minutes). So the full name indicates that the source was found in the Ohio survey at Right Ascension 8h 53m and Declination +25 ± 5°.

5. The term "Lacertids" has now disappeared from the astronomical lexicon, although it can still be found in some older texts, although it can be confused with Lacertid meteors.

6. BL Lac had been discovered as a variable star in 1927 and observed as such for more than 60 years before anyone realized that it was not a normal star in our Galaxy. It was only when it was realized that the position of a quite strong radio source with an unusual spectrum coincided with this variable star in the constellation of Lacerta that astronomers decided to investigate further. What they found was that the "star" had a totally unstarlike spectrum with some weak absorption lines showing that it had a red shift of 0.069, putting it at a distance of 900 million light years. What is more, when it was faint—it could get as bright as magnitude 13, but at times faded down to magnitude 17 and fainter—a giant elliptical galaxy was clearly visible, with the quasar in the center. This was the first clear proof that quasars occur in the center of galaxies. The term *BL Lac* objects was coined to describe quasars that showed blank or almost blank spectra and that were extremely variable, often changing in brightness by a magnitude in a few days; if this does not sound like much, think of it in terms of being equivalent to turning on or off in this time as many stars as in ten galaxies like the Milky Way.

7. R. D. Wolstencroft, G. Gilmore, and P. M. Williams, "Rapid Variability of OJ 287 at 1.25 Micron," *Monthly Notices of the Royal Astronomical Society* 201 (1982): 479–85. This is such bizarre and improbable behavior for such a remote and luminous object that the astronomers involved from the Royal Observatory Edinburgh are, to this day, both skeptical about what they saw and, at the same time, unable to explain it as other than a real variation of OJ287. Similarly, other astronomers are skeptical but can see no other obvious explanation. This particular observation has entered the folklore of the observation of quasars as one of the most peculiar and inexplicable observations ever made.

8. Two research papers published in April 1984 both used the term in their title and thus seem to have priority as having been the source of the term blazar: R. R. J. Antonucci and J. S. Ulvestad, "Blazars Can Have Double Radio Sources," *Nature* 308 (1984): 617–19; and J. F. C. Wardle, R. L. Moore, and J. R. P. Angel, "The Radio Morphology of Blazars and Relationships to Optical Polarization and to Normal Radio Galaxies," *Astrophysical Journal* 279 (1984): 93–98, 101–11.

9. N. Visvanathan and J. L. Elliot, "Variations of the Radio Source OJ 287 at Optical Wavelengths," *Astrophysical Journal* 179 (1973): 721–30.

10. Alfred Frohlich, Shmuel Goldsmith, and Donna Weistrop, "Further Studies of the Optical Variability of OJ 287," *Monthly Notices of the Royal Astronomical Society* 168 (1974): 417–26.

11. Alan L. Kiplinger, "On the Short-Timescale Variability of OJ 287," *Astrophysical Journal* 191 (1974): L109–10.

12. Esko Valtaoja is one of the greatest characters of world astronomy. His long, bushy beard and large build gives him an extraordinary resemblance to the members of the American rock group ZZ Top—some astronomers even know him by this name! One of the things that most endears Esko is that when you enter his office at Tuorla Observatory (near Turku, Finland) the first thing that you see is a large photo on the wall of him, completely naked after leaving the sauna, about to jump through a hole in the ice to have a nice, refreshing swim to cool off and close the pores.

13. E. Valtaoja, T. Korhonen, M. Valtonen, H. Lehto, P. Teerikorpi, H. Terasranta, E. Salonen, S. Urpo, M. Tiuri, and V. Piirola, "A 15.7-min Periodicity in OJ287," *Nature* 314 (1985): 148–49; L. Carrasco, D. Dultzin-Hacyan, and I. Cruz-Gonzalez, "Periodicity in the BL Lac Object OJ287," *Nature* 314 (1985): 146–48.

14. For what it is worth, my student and I seemed to see periodic behavior in observations made with the 2.5-meter Isaac Newton Telescope in La Palma in March–April 1987, but the period was of 19 minutes, exactly in between the period seen by Valtaoja and collaborators and that seen by Carrasco and collaborators (José Antonio de Diego and Mark Kidger, "A Possible Nineteen-Minute Periodicity in the Light Curve of OJ 287," *Astrophysics and Space Science* 171 [1989]: 97–104). In 1986 William Kinzel and colleagues at the University of Massachusetts saw a 35-minute period in radio observations (W. M. Kinzel, R. L. Dickman, and C. R. Predmore, "A Possible 35-Minute Periodicity in the OJ 287 active Galactic Nucleus at 7-mm Wavelength," *Nature* 331 [1988]: 48–50). But observers at the Very Large Array (VLA) observing simultaneously with Esko Valtaoja and his colleagues saw no evidence of periodicity whatsoever in their radio observations (J. W. Dreher, D. H. Roberts, and J. Lehar, "Very Large Array Observations of Rapid Non-Periodic Variations in OJ 287," *Nature* 320 [1986]: 239–242). And even Kinzel and collaborators saw nothing unusual in the radio light curve of OJ287 when they looked again a year later.

15. Some years later on a visit to Mexico I was able to discuss the results with Luis Carrasco and on a separate visit to Finland I spoke to Esko Valtaoja and, on different occasions to Harry Lehto, one of his coauthors and someone who has been for many years a good friend, about these differing results and their meaning. All were uncertain about what had happened, although Harry did comment that one of the things that troubled him was that the period in the Finnish radio data seemed to come and go at will over the months that they were observing, which made him feel that it was not really periodic behavior at all.

In one particular conference publication that I prepared in the 1990s, I took all the different periodicities that I could find in the literature and examined them. My particular conclusion was that all of the periods reported seemed to be close to either 19 minutes or 38 minutes (exactly twice 19 minutes), with nothing in between. This suggests that there is a special significance to this period, but it is not obvious what it might be, given that it is not now thought that we can see the black hole directly in blazars. An idea of the problems posed by OJ287 and its "periodicities"

can be obtained by reading a brief presentation that I made to the American Astronomical Society in 1992 (Mark Kidger, José A. de Diego, Leo Takalo, Kari Nilsson, Merja Tornikoski, Aimo Sillanpaa, and Filippo Zerbi, "Periodicity of OJ287 in Multisite Multifrequency Fast Photometry? . . . Now You See It and Now You Don't," *Bulletin of the American Astronomical Society* 181 [1992]: 1102; available online at http://cdsads.u-strasbg.fr/cgi-bin/nph-iarticle_query?1992AAS . . . 181.1102K&data_type=PDF_HIGH&type=PRINTER&filetype=.pdf).

16. A. Sillanpaa, S. Haarala, M. J. Valtonen, B. Sundelius, and G. G. Byrd, "OJ 287-Binary Pair of Supermassive Black Holes," *Astrophysical Journal* (1988): 325, 628–34. Scientists measure how much importance, or *impact,* a scientific article has had by counting how many other studies published afterward make reference to it. This particular article has been cited more than 150 times, making it an important and very well-known study.

17. It usually takes a minimum of about nine months and often more than a year for a scientific article to go through the complex process of refereeing (that is, scrutiny by other scientists), be revised and re-revised, and finally go through the different stages of preparation of the definitive manuscript and publication. This means that results may sometimes, in this risky world of prediction, be out of date before they ever see the light of day. But, of course, nothing ventured, nothing gained.

I myself had carried out a detailed study looking at the many predictions made about the blazar 3C345 (M. Kidger, "The Optical Variability of 3C 345," *Astronomy and Astrophysics* 226 [1989]: 9–22), and throughout the 1960s, 1970s, and 1980s there was a long controversy about possible periodicity in the quasar 3C273 in which different groups debated, sometimes heatedly, the validity or otherwise of possible periods. My own conclusion, from this and other studies, was that there is something in quasars that does permit short-term behavior that looks rhythmic and resembles periodicity, but that these are unstable and tend to break down quickly. This sort of behavior I denominated *pseudo-periodicity*—in other words, it might *look* for a time that something was happening in regular fashion, but it was not really, and the regularity was just an illusion.

In all, different groups must have suggested that some kind of periodicity was present in at least a dozen quasars, but none of these studies was particularly well received. By the time that it had become evident that it was just not possible to predict the behavior of quasars, most astronomers had developed such a strong aversion to the idea of any kind of periodic behavior that it was difficult to get them to take it seriously at all in any way, shape, or form.

18. M. Kidger, L. Takalo, and A. Sillanpaa, "A New Analysis of the 11-Year Period in OJ287—Confirmation of Its Existence," *Astronomy and Astrophysics* 264 (1992): 32–36. This paper was fairly well received, but the fact that even after a century we cannot give a better value of the period than this has always worried me. After so many years, $\pm$ 0.5 years adds up to a pretty big uncertainty—it means that 10 outbursts ago we can fix the time of the outburst to only $\pm$ 5 years, which is not really very good. We will come back to this issue.

19. What was not so widely mentioned is that it is a former leper hospital and that one of the very few sights to visit on Seili Island is the cemetery. In summer the research station seems to be almost entirely populated by mosquitoes who evidently spend their summer holidays there en masse (with 80,000 lakes, Finland is not short of a mosquito or two . . . million).

20. A CCD camera is an electronic camera. CCD stands for "Charge Coupled Device." CCDs are now familiar to millions because they are used in digital cameras and in video cameras. A CCD uses a special light-sensitive electronic chip made of silicon that converts the light that falls on it into tiny electrical signals. These can then be processed and analyzed and converted into an image by a computer.

21. Two details give an idea of the standard of Paul Boltwood's work. Observing from his city site with a small telescope and using standard filters, he was regularly measuring a magnitude 20 blazar for us and obtained usable data—no mean feat. Later he installed a 60-centimeter telescope with which he won a competition organized by the magazine *Sky & Telescope* for the faintest object detectable by an amateur observer; Paul's winning exposure reached magnitude 24, which is fainter than the famous 5-meter (200-inch) telescope at Mount Palomar had been able to reach before the invention of the CCD camera.

22. This is a facility whereby 5 percent of all the time on the international telescopes in the Canary Islands is reserved for very large and ambitious projects that would otherwise not get time. This means using all the different telescopes and supposes a huge amount of telescope time. Over the years I have participated in three such projects, leading two of them.

23. Be patient and watch this space. This is the longest range prediction that I have ever made. We only have another 25 years to wait to see if it is correct.

24. Gary is big, but he is not that big!

25. Mark R. Kidger, "The 11 Year Period in OJ 287 Revisited: Is It a True Long-Enduring Period?" *Astronomical Journal* 119 (2000): 2053–59.

26. The official translation and explanation seem to suggest that whoever coined this name did not speak Spanish very well.

27. Their orbits reached an altitude of 100,000 kilometers, or a quarter of the distance to the Moon, well above geocentric altitude, meaning that the satellites' lifetime would be effectively unlimited against atmospheric decay.

28. You can find a news report on this theory at http://news.bbc.co.uk/1/hi/sci/tech/4433963.stm.

29. Astronomers class objects with polarized light as being of one of two types: *intrinsically* polarized objects, that is, objects that produce polarized light because of the way that their energy is generated, which includes quasars, supernovae, and other extremely violent objects; and *extrinsically* polarized objects, in which the light emitted from the object is not polarized, but it is then reflected off something like a nebula around the star that polarizes it. Only in intrinsically polarized objects do we learn about the object itself from observing its polarization; in extrinsically polarized objects, we only learn about the nebula that is reflecting the light.

30. Astronomers, and especially cosmologists, can be extremely cynical and sometimes argue that an object that is as close to us as 2.5 billion light years is so nearby that it is barely worth bothering with in normal circumstances. Such astronomers also tend to regard the Moon as a regrettable aberration and a nuisance that is of no use or interest whatsoever.

CHAPTER 4. How Far Is It to the Stars and Will We Ever Be Able to Travel to Them?

1. It is interesting to speculate how history might have been different if these two great civilizations had used a number system that made mathematics practical. Some people will have struggled with the intricacies of doing simple sums with Roman numbers in school—try adding even two simple numbers like LXVIII + XCVII (68 + 97), and you will rapidly understand why multiplication and division with Roman numerals was, if not impossible, so much hassle that no sensible person would try it. The Roman system, though, was positively simple and streamlined compared to the Greek system of using a different letter for each number and a dot over the top of the letter to signify a power of ten. One line of speculation is that the legend of Atlantis came about in part because Plato became confused with his numbers and stated that Atlantis was 10 times bigger than it really was. Because of the near impossibility of doing anything other than the simplest arithmetic, the Greeks and later the Romans specialized in philosophy and geometry, which did not require mathematics. Had the Romans been able to combine their brilliant engineering skills with mathematical analysis, they could possibly have advanced their technology to as high as a twentieth-century level with unimaginable consequences for the history of the world. It was not until the Arabic numbers that almost everyone uses now were adopted in the Western world that the science of mathematics really started to take off.

2. The calculation is extremely simple. Say that the difference in angle—the height of the Sun in the sky—between two points is 5 degrees. A full circle is 360 degrees, which is 72 times greater. Thus, if the two points are separated by 550 kilometers, the circumference of the Earth is 72 × 550 = 39,600 kilometers.

3. Astonishingly, some mealymouthed critics have pointed out that if the larger value is correct, his calculation was not so good. Even the "bad" value was just 4 percent out—1 part in 25—which is amazing accuracy. Given that Eratosthenes rounded his answer to one part in 25(!), that is, to the nearest 10,000 stadia, it is as good as he could possibly have done without giving extra figures of accuracy.

4. Some people have argued that Galileo did not exactly help himself by being highly undiplomatic in the way that he expounded this theories in public and that it was at least in part his own fault that his close friend, the pope, intervened. Even so, the fact that it took more than 300 years for Galileo to be pardoned seems excessive.

5. It is a fortunate historical accident that Kepler chose to work with the orbit of Mars, a planet with a very elliptical orbit compared to that of most of the planets. Had he chosen to work with Tycho's observations of Venus, he would have been hard put to detect the tiny deviation from being a perfect circle.

6. Bradley was a remarkable man. Born in 1693, he went to Oxford University and then became the vicar of the village of Bridstow, in Monmouthshire, on the border between Wales and England. At the young age of 28 he became professor of astronomy at Oxford and in 1742 succeeded Edmond Halley as the third Astronomer Royal.

7. How much is 710,000? An American one-cent coin is almost 2 centimeters across, so a line

of 710,000 of them would be about 14 kilometers long. That may help you to imagine this number better.

8. Herschel described the Galaxy as looking like "a cloven grindstone."

9. The island of Tenerife, where I lived for many years, has a very enthusiastic UFO community and is the UFO capital of Spain. To my permanent bemusement, one of the basic tenets of faith of this community is the fact that there is a UFO base at a place on the south coast of Tenerife called Montaña Roja (Red Mountain), an extinct volcano. The most unusual thing about this theory is the fact that the mountain is at end of the runway of a major international airport with its corresponding radar. Various friends of mine work in air traffic control at the airport, and I can only wonder how they cope with so many interstellar takeoffs and landings in addition to all the domestic and international flights that they must control. There is also, apparently, a great deal of UFO activity around Mount Teide in the center of the island. Having spent more than 1,000 nights at the observatory close to Mount Teide, and many hours outside looking at the spectacular sky from in and around the observatory without ever seeing a UFO, I can only wonder what my fellow professionals and I are doing to earn our pay up there if we cannot see these UFOs ourselves. A more skeptical point of view is that the beaches in the south of Tenerife, around the airport, are a popular landing spot for smugglers who wish to bring contraband—particularly tobacco—ashore. The observation of strange lights on these beaches at dead of night is thus likely to be of more concern to the police than to extraterrestrial-minded earthlings.

10. Of course, various of Arthur C. Clarke's novels and short stories are based around interstellar travel, for example, *2001: A Space Odyssey*, *Rendezvous with Rama*, *Childhood's End*, *Rescue Mission*, and *The City and the Stars*. One assumes that he has since recanted, or classes these novels as science fantasy, where everything is permitted, quite apart from some of his wonderful comic short stories where all is possible, even extraterrestrials confused into thinking that Donald Duck is an average inhabitant of planet Earth. Very few writers have ever written good science fiction humor, and these short stories are some of the very best of their class.

11. It is a salutary, though somewhat nauseating thought that in each glass of water that we drink there are some seven molecules that passed through the kidneys of Julius Caesar. On Earth we try not to think of such recycling. In a generation ship, it would be essential to do the same and for nothing to go to waste.

12. These remain two of my favorite science fiction stories of all time, although they were written in 1950 and 1956, respectively. Although they make an extrapolation of technology that is of uncertain practicability (the conversion of large amounts of matter directly to energy), there is little in them (apart from the application of telepathy for instant communication over many light years) that even 50 years later one can definitely say is not possible. Interestingly, it is the earlier of the two books (*Farmer in the Sky*) that has the most consistent and plausible science, a remarkable achievement for a book set at least a century in our future.

13. One problem, in particular, is that we make antimatter atom by atom and that no one has yet made enough even to be seen with a microscope, let alone in the multimegatonne quantities

that the *Enterprise* would need. There are sound laws of particle physics that *seem* to prohibit ever making antimatter in large quantities.

14. You can find this report at http://news.bbc.co.uk/1/hi/sci/tech/4564477.stm.

15. You can read a summary of this, including a short interview with the author of the study, in the BBC's excellent science and technology Web pages at http://news.bbc.co.uk/1/hi/sci/tech/364496.stm.

16. Given the lack of space on board and the total lack of amenities, it is hard to imagine that these interstellar trips can last for more than a few hours. Luke's ship also seems to break all records for fuel economy, with a quite spectacular number of "miles to the gallon."

CHAPTER 5. How Old Is the Universe?

1. This figure is much less than the known age of the Earth. The reason is one that would not be understood for some 70 years. In fact, like the continents, the oceans are constantly being born and dying. The Atlantic Ocean itself only opened about 180 million years ago; although it was a landlocked inland sea for tens of millions of years. South America and Africa only split around 75 million years ago, opening the young North Atlantic Sea to the world's oceans. Thus the Atlantic Ocean is itself much younger than the age of the Earth.

2. The way this works is that a radioactive atom decays by throwing off an alpha particle (a helium nucleus) or a beta particle (an electron), which may be accompanied by a gamma ray. The particle that is thrown off by the nucleus has kinetic energy—that is, energy by virtue of its mass and velocity (the kinetic energy is calculated by multiplying half the mass by the square of the velocity or, mathematically, $E = \frac{1}{2} mv^2$—hence, the greater the mass and the velocity the greater the energy). This energy is absorbed when the particle is stopped by colliding with other atoms, thus heating the substance. If one is unwise enough to pick up a piece of uranium or plutonium and hold it, you will notice that it is warm to the touch—this is your skin absorbing the energy of the radioactivity.

3. Marie Curie's career is no less remarkable for being so well known. In 1911 she added the Nobel Prize for Chemistry, won jointly with her husband, to her Nobel Prize for Physics. Sadly, the dangers of working with radioactivity were unknown at the time, and she died largely due to her constant exposure to unhealthy levels of radiation.

4. In fact, this is not *quite* true. Some tiny diamonds have actually been found in a meteorite that are older still and that formed around another star before being incorporated into the material from which our solar system formed.

5. If a star has more heavy elements in it, the spectral lines from these elements are stronger. These lines are called *absorption lines* because they absorb the light of the star and are *dark* in the spectrum. That means that the more lines that it has in its spectrum, the dimmer the star appears to be, so we think that it is further away than it really is.

6. B. Chaboyer, P. Demarque, P. J. Kernan, and L. M. Krauss, "The Age of Globular Clusters in Light of Hipparcos—Resolving the Age Problem," *Astrophysical Journal* 494 (1998): 96–110.

7. Note, though, that there is still an uncertainty of about 10 percent in this value. Statistically, what this means is that there is about a one in three chance that this age is as much as 10 percent off—either 10 percent older or younger. Still, an astronomer would say that 1,000 million years between friends is nothing (this is illustrative of the sorts of problems that astronomers face in answering this kind of question where, despite every conceivable effort, the results are often not as good as we would like—the Universe rarely permits absolutely definitive answers about anything).

8. This subject is also tackled in "Pluto: Impostor or King of the Outer Darkness," chap. 7, in Mark Kidger, *Astronomical Enigmas: Life on Mars, the Star of Bethlehem and Other Milky Way Mysteries* (Baltimore: Johns Hopkins University Press, 2005).

9. This presentation was later published in the magazine *Popular Astronomy*. See V. M. Slipher, "Spectrographic Observations of Nebulae," *Popular Astronomy* 23 (1915): 21–24.

10. In this paper Slipher added at the end that he had detected that one of the galaxies was rotating—of itself a fundamental discovery.

11. Slipher also suggested in this paper that the Milky Way shows what we now call "a peculiar velocity"—that is, that it is falling to a certain point in space to which the gravitational attraction of other galaxies is pulling it. Although his measure of the direction of movement was wrong, mainly due to the fact that most of his galaxies were in two small regions of the sky, his estimate of the size of the peculiar velocity was actually correct to within a factor of about two. Again, though he did not realize it at the time—this discovery was to be delayed a number of years—his observations of the galaxy NGC 1068 in Cetus showed that it is an active galaxy or, in the modern argot, an AGN. This galaxy showed not only the dark lines of stars in its spectrum, but also the bright lines that we now recognize as symptomatic of violent activity in its nucleus. As this galaxy was the first object that I ever observed as a professional astronomer (in a program of infrared observations of AGNs with the 1.5-meter Carlos Sánchez Telescope in Teide Observatory on October 21, 1984), I find great satisfaction in reading of Slipher's discovery of its nature.

12. K. Lundmark, "The Determination of the Curvature of Space-Time in de Sitter's World," *Monthly Notices of the Royal Astronomical Society* 84 (1924): 747–70.

13. Remarkably, he also took literature and Spanish.

14. Strictly speaking the Birr Castle 72-inch (1.8-meter) telescope was still the second largest in the world, but this telescope was only used visually, thus greatly underusing its power, and, by this time, had not been used for some years.

15. Hubble was evidently proud of his military service, as he returned to Mount Wilson in full uniform and used his rank as an introduction with all the people he met. When the United States entered World War II, he went to the Aberdeen Proving Ground in Maryland where he signed on as head of ballistics. The *Time* biography comments that on one occasion he spent an entire afternoon firing round after round with bazookas "at great personal risk" in an attempt to detect a design flaw.

16. Later, in what was an unusual move that would certainly be frowned upon these days, Hubble was to hire a publicist in an attempt to gain the Nobel Prize that he apparently craved, although he was already a major public figure, courted by Hollywood stars. By the time that the Nobel Prize

committee was prepared to accept astronomy as a branch of physics and thus eligible for the prize, it was too late. He died in 1953, at the relatively young age of 63, having already a few years previously suffered a severe heart attack, just at the time when it appears that the committee was finally considering him strongly for this ultimate honor in science.

17. If we were to be nitpicking, we would say that both were wrong, although both had their conclusions correct. Shapley had greatly overestimated the size of the Galaxy with his original value of 300,000 light years (the currently accepted value is 100,000), and Hubble had greatly underestimated the distance to the Andromeda Galaxy (he estimated 900,000 light years and the correct value is 2.2 million). So, rather than being a little way outside the Milky Way, the Andromeda Galaxy was at a distance corresponding to more than 20 times the size of our Galaxy.

18. In 1931 Einstein visited Hubble at Mount Wilson and reputedly thanked him personally for saving him from his own folly. Recently, though, cosmologists have found the need to reintroduce this term as it seems that there really is an antigravity, inflationary force that is counteracting the effects of gravity, so one now wonders which was the true blunder. Curiously, Einstein had originally believed his conclusion that the Universe was expanding and was persuaded by astronomers of the day that it was really stable, thus leading him to modify the equations and add the Lambda term to them. Quite what Einstein would have made of modern results that require the Lambda term to be real and quite substantial one can only guess; one suspects that he would have been reluctant to reverse himself on the issue a second time.

19. It is interesting, in view of previous comments, that, despite including results from his assistant, Milton Humason in this paper, it was published under his own name only. The later papers presenting further results were all coauthored with Humason.

20. It has been noted cynically that the Hubble constant is possibly the most variable constant in the history of astronomy with the different estimates of it made over the years varying by a factor of more than 20.

21. All this should not have been a complete surprise. As early as 1933 the Dutch astronomer Jan Oort had cast doubt on Hubble's distances and suggested that they were underestimated by a factor of about two. However, this work was largely ignored at the time and is today almost forgotten.

22. This was about the time of the *Challenger* accident, which caused a three-year delay in the launch of the Hubble Space Telescope.

23. You can find a detailed report on the WMAP satellite and its results at www.space.com/scienceastronomy/map_discovery_030211.html.

CHAPTER 6. Is Anybody There?

1. Oddly, in the original novel by H. G. Wells, the Martians, although far superior militarily, were far from invulnerable and suffered significant losses in the fighting before their final defeat. The cinematographic versions of the story, both in the 1950s and the more recent remake, added the invulnerable force shield to their armory and made them genuinely unbeatable.

2. An interesting case is computer circuitry and electronics, which are, of course, silicon based.

At present, few experts would argue that even complex electronic devices are, in any sense, alive, but that may not always be the case. What about HAL9000, the famous computer from *2001: A Space Odyssey*? There it is less easy to say. Could it be that one day we will discover that the ambassador from Tau Ceti causes us severe protocol problems because he looks like a laptop computer?

3. See "Goldilocks and the Three Planets," chap. 10 in Mark Kidger, *Astronomical Enigmas: Life on Mars, the Star of Bethlehem and Other Milky Way Mysteries* (Baltimore: Johns Hopkins University Press, 2005).

4. "Our Moon: Nearest Neighbor—Hot Property?" chap. 5, in ibid.

5. "Are We Stardust?" chap. 12, in ibid.

6. "Is There Life on Mars," chap. 6, in ibid.

7. You can find a report on this news item at http://news.bbc.co.uk/1/hi/sci/tech/2282168.stm. A summary of the original report in the September 28, 2002, edition of the magazine *New Scientist* is given at www.newscientist.com/hottopics/astrobiology/venusmicrobe.jsp.

8. Methane is a gas that is somewhat unstable and broken down by ultraviolet light. This means that unless there is an internal source of methane to replenish what is lost, the methane should have disappeared from the atmosphere long ago. One possible source of methane is volcanoes; another, more exotic possibility, is that the source of methane (as on Earth it is) at least in part, from bacteria deep below the surface. See http://news.bbc.co.uk/1/hi/sci/tech/4196261.stm for more details.

9. "Goldilocks and the Three Planets," chap. 10, in Kidger, *Astronomical Enigmas.*

10. "Are We Stardust?" chap. 12, in ibid.

11. Later the name was changed to SETI, pronounced the same, but with "communication" changed to "search" to reflect the fact that we can only listen as interstellar distances are so great that two-way conversations are impossible.

12. Very unusually, the publicity-conscious NASA refused this presidential request, something that should indicate to the reader that the agency felt that it had no significant possibility of finding anything worthwhile in such a study.

CHAPTER 7. How Will the Universe End?

1. To my amazement, I have found that there is even a cosmologist called James Bond who works on these problems!

2. It is not true to say that you "escape from the Earth's gravity." You do not. The Earth's gravity is still there, it is still influencing you, but it is powerless to stop your flight. It is more like a running back who weighs 230 pounds: even though two members of the defense are hanging on to him and trying to stop him, he still drags them into the end zone and scores.

3. The current age of the Universe would be two-thirds of the Hubble time. In other words, if the Hubble constant is 70 km/s/Mpc, the Hubble time—the age of the Universe if the expansion is not slowing—would be 14 Gyr, and for a just critical Universe with $q = 0.5$, its real age would be 9.3 Gyr.

4. Reber had received a bachelor's degree from the Illinois Institute of Technology in 1933 and,

in 1937, built his own radio telescope in his back yard. Reber's was the first radio telescope of the modern dish design.

5. Scientists involved in the British war effort regard this register and its effective use as having been one of the most important factors that allowed Britain to compete so successfully with the Nazi scientific war machine, despite the Nazis having had several years' head start.

6. Few people really understand the importance that radar had in the history of the Battle of Britain. In 1937 the Luftwaffe had sent the airship *Hindenburg* on a mission to fly along the south coast of England to look out for telltale radio emissions that would let the Germans know that the Royal Air Force (RAF) was using radar to detect enemy aircraft. Despite the fact that the Women's Auxiliary Air Force (WAAF) officers tracked it all the way and had never seen such a large signal, the scientists on board heard nothing—why they did not is a major mystery. This convinced the German high command that the RAF had nothing better than visual observation and sound detectors to track incoming aircraft and would thus have to maintain standing patrols over southern England that would be ruinously expensive in men, aircraft, and scarce petrol. When the Luftwaffe began its massive attacks, the WAAFs would see them massing across northern France and then track them across the English Channel, allowing the controllers to first alert the RAF squadrons that were in the likely target area and then scramble them and guide them to an interception. The Luftwaffe was disconcerted to discover that, whenever their bombers approached a target, there were usually RAF fighters waiting for them. Although the official statistics produced after the heat of battle had died down showed that the Battle of Britain was more a bloody stalemate than the glorious victory that war films usually portray, the Luftwaffe's aim of destroying the RAF and dominating the skies over the British Isles was thwarted, largely thanks to radar and, as a result, I and many other children like me grew up speaking English and not German.

7. The explanation is that these are active galaxies with a giant black hole in the center. In these objects, gas is falling onto the central black hole in the galaxy. Sometimes, though, more gas starts to spiral in than the black hole can swallow at any one moment. The result is that the extra gas is blown off from the poles of the black hole in exactly opposite directions. This jet of gas is full of electrons moving at very high velocity and spiraling in magnetic fields, which emit radio— what is called synchrotron radiation—very vigorously, but it is far too thin and faint to be seen visually. What was seen in the radio observations was the giant cloud at each end of the jet where the gas starts to disperse into intergalactic space. Ordinary galaxies, though, emit very little radio energy and can be detected only if they are very close like the Andromeda Galaxy.

8. This spurred a sort of international race between different groups to try to measure smaller and smaller objects using telescopes separated by ever greater distances until, in the early 1960s, even with astronomers combining telescopes in Europe and the United States, separated by thousands of kilometers, a few objects such as 3C345 still resisted stubbornly, showing that they were smaller than 0.0004 arcseconds in diameter. Astronomers were, of course, interested to see what kind of object could be so tiny and made considerable attempts to identify them with optical telescopes. The result was the discovery of quasars with 3C345 being one of the first to be identified.

9. When Marvin the Paranoid Android moans at one point in the original *Hitchhiker's Guide to the Galaxy* radio series, "Oh dear! Reality is on the blink again," one suspects that he is a frustrated cosmologist who has just found out that the Universe is being difficult and that yet another theory to explain it has failed.

10. Strictly speaking the name is q_0, which is the value of q at this instant in time.

11. Some important inconsistencies with this theory worry astronomers. If this theory is correct, the supernova explosion will take place alongside a giant star that is close enough so that its mass has been falling on the white dwarf. The explosion should blow huge quantities of hydrogen gas off the outer layers of the giant star. This hydrogen should mix with the gas blown off by the supernova and produce an expanding gas cloud rich in hydrogen. However, this is not seen. In fact, just one type Ia supernova rich in hydrogen has been seen, which makes one suspect that something is seriously wrong. There is an alternative theory that a type Ia supernova may be due to the merging of two white dwarf stars, although in this case it is less easy to see just why the explosion should always be the same size. The important thing about using type Ia supernovae as standard candles is that the method seems to work, whatever the explanation.

12. Again, there is a worrying little doubt about some type Ia supernovae that may be too faint, but the overwhelming majority do seem to give excellent results. Nothing in life is ever perfect, although type Ia supernovae seem to come close enough for most purposes.

13. The International Astronomical Union (IAU), the governing body of astronomy, recognizes 3,182 supernova discoveries in history up to June 2, 2005 (the date that this chapter was first finished), with the first being the brilliant 1006 supernova discovered by Chinese astronomers in Lupus. Of these, 1,492—almost half of the entire total—have been discovered just since January 1, 2000. Only 752 supernovae were discovered before 1990.

14. Taking 30-second exposures, this telescope reaches magnitude 19.5 and can take 80 to 90 images of galaxies each hour.

15. If you don't believe me, go to http://cfa-www.harvard.edu/cfa/oir/Research/supernova/HighZ.html and select the menu "Astronomers" to find information and names of the supernovae. The team proves that astronomy and a sense of humor are compatible and that even the most serious and fundamental project can be fun.

16. You can find the list of supernovae and their magnitudes at http://cfa-www.harvard.edu/iauc/07300/07312.html.

17. This was the famous "Hubble Deep Field." The idea was to select two small areas of the sky and observe them constantly for 10 consecutive days to observe the faintest objects ever detected and thus to see what there was in an apparently completely blank area of sky. This has allowed the Hubble Space Telescope team to see the faintest and most distant galaxies ever detected in the Universe.

18. You can find the discovery announcement and more details of the two supernovae at http://cfa-www.harvard.edu/iauc/06800/06810.html. An explanation of how Sn 1997ff was studied using images taken quite accidentally by other Hubble Space Telescope observers who had no

idea that the supernova was there can be found at www.lbl.gov/Science-Articles/Research
-Review/Magazine/2001/Fall/departments/frontline/supernova.html.

CHAPTER 8. Why Is the Sky Dark at Night?

1. Little did I imagine that something like this event would later be repeated. In January 2007
astonished astronomers in the Northern Hemisphere saw how the tail of Comet McNaught pro-
jected over the horizon in exactly the same way as Comet de Chéseaux had in 1744. It is now evident that the spectacular appearance of de Chéseaux's comet had the same explanation. You can
see some images of the two comets and an explanation at www.astrosurf.com/comets/cometas/
2006p1/Analysis/McNaught_DeCheseaux2%20(2).htm.

2. "Goldilocks and the Three Planets," chap. 10, in Mark Kidger, *Astronomical Enigmas: Life on
Mars, the Star of Bethlehem and Other Milky Way Mysteries* (Baltimore: Johns Hopkins University Press,
2005).

3. The green color comes from the emission of light by ionized oxygen. This can be strong
enough in some cases to be detected with even a small telescope.

4. In *The Scenery of the Heavens* (London: Roper & Drowley, 1890), Gore, a Fellow of the Royal
Astronomical Society, describes his book as "a popular but exact description of the most interesting facts relating to the planets, comets, meteors, fixed stars and nebulae."

5. "Pluto: Imposter or King of the Outer Darkness?" chap. 7, in Kidger, *Astronomical Enigmas.*

6. For many years it was assumed that the cosmological constant was zero. However, 75 years
later research into the expansion of the Universe has shown that the cosmological constant is almost certainly not zero. Distant galaxies seem to be receding faster than can be understood unless there is a cosmic repulsion force that is partially counteracting the slowing of the expansion
caused by gravity.

7. Robert Dicke's work was not recognized by the Nobel Prize committee despite his insight being necessary for Penzias and Wilson to recognize what they had detected. Some astrophysicists still feel that this was a scandalous decision, others feel that it was Penzias and Wilson
who had done the hard (and dirty) work and that they alone deserved the recognition. Like many
times in science both views have their merit. You the reader can make up your own mind about this.

8. Two groups made the discovery virtually simultaneously. Most books attribute the discovery to the COBE satellite alone, although this is somewhat of an injustice. From 1984 a group of
British and Spanish researchers from Jodrell Bank, Cambridge University, and the Instituto de
Astrofísica de Canarias had been searching for anisotropies in the CMB using an instrument at
Teide Observatory in Tenerife. After six years of collecting data on a shoestring budget, one point
on the sky started to show a small bump in the signal that was steadily growing in significance as
more data were added. Early in 1991 the group was ready to announce this result as a possible first
anisotropy in the CMB and wrote a letter to be published in the prestigious journal *Nature*. Before
the letter could be published, however, the COBE team announced its results in Washington in a
blaze of publicity in April 1991. When the COBE data were compared with the data from Teide,

it was confirmed that the same signal from the same point of the sky was present in both data sets. The happy ending to this story is that the two groups are collaborating together on the next satellite that will be launched to study the CMB, the Planck satellite, which will be launched in 2008 from the European Space Agency's spaceport at Kourou in French Guyana.

CHAPTER 9. How Do We Know There Was a Big Bang?

1. In fairness to Canute, who tends to get a poor press from more history-challenged people, he was trying to make a point. His courtiers were convinced that he was so powerful that he could stop the very waves. Canute had no such illusions and ordered that his throne be set up on the shore to show them otherwise. He sat and, as the tide came in, commanded the waves to stop (which, or course, he knew that they would not). As the waves started to lap around him, the selfsame courtiers begged him to come back and not to risk drowning. Canute persisted until even the most stubborn of his adoring entourage could see that he did not have the powers that they had attributed to him. Rather than praise King Canute for his common sense, many people lampoon him thinking that it was he himself who thought that he could defy the waves.

2. Note that the early idea after the conversion of mass to energy was "discovered" by Einstein was that a star like the Sun would last as long as it took to convert its enormous mass of 2×10^{30} kilograms (that is, a "2" followed by 30 zeros) into energy at a rate of 4 billion kilograms per second, so the Sun would continue to shine for about 100 trillion years. It was only later that it was discovered that just a small fraction of the Sun's mass ever gets converted into energy and that the Sun will live for "only" 10 billion years, although smaller stars will endure many times longer.

3. Arp worked as an assistant to Edwin Hubble toward the end of Hubble's career.

4. Astronomers regularly detect such lateral motion when looking at radio loud quasars. If there is a jet of gas leaving the quasar, radio astronomy techniques are often able to detect clumps of material moving along the jet.

5. Often science cannot offer a definitive answer; it can only offer a probability that a particular theory is correct or not. In other words, all too often it is impossible to say that one theory is right and another wrong, we can only say that one is *probably* right and the other *probably* wrong. Statistics in science is just like playing the lottery—even if your chances of success are tiny, someone (almost) always wins. A 1 percent probability may sound small, but play often enough and that 1 percent probability will come up in the end, so scientists will usually say that something that has a probability of 99 percent is only barely significant.

6. A paper by him in 2001 reiterated his view that there are noncosmological red shifts. See G. Burbidge, "Non-Cosmological Red Shifts," *Publications of the Astronomical Society of the Pacific* 113 (2001): 899–902. You can find this paper at www.journals.uchicago.edu/PASP/journal/ issues/v113n786/201131/201131.html.

7. Among the discoveries that convinced all but the most diehard opponents of the cosmological red shift were the following: the large number of quasars that showed absorption lines in their spectra due to their light being absorbed by more nearby galaxies with smaller red shift, the increasing number of cases where a faint underlying galaxy could be detected with the quasar in

its center, and the increasing number of cases where a quasar was found surrounded by faint galaxies at the same red shift showing that quasars are associated with clusters of galaxies.

8. For the pedant, of the first 92 elements in the periodic table of elements, one, technetium, as its name implies, does not occur naturally on Earth, although it has been made in nuclear reactors and *does* occur naturally in red giant stars where many unstable isotopes of elements are formed. In contrast, neptunium and plutonium, the first two transuranic elements *do* occur naturally on Earth, although only in tiny quantities in uranium-rich rocks. Much heavier elements than plutonium are formed in supernova explosions and, although only artificially created on Earth, do exist, if only briefly, in space.

9. This delightful phrase was used by Jocelyn Bell-Burnell, the discoverer of pulsars, to describe the conditions in the interior of a neutron star in which all the atoms are crushed into a dense mix of elementary particles with names that are traditionally taken (until the number of particles discovered outstripped Greek's capability to supply new letters) from letters of the Greek alphabet like lambda, pi, sigma, etc. In other words, a rich Greek alphabet soup is a fantastically dense state of matter where there are no atoms, just huge numbers of elementary particles of different kinds.

10. As in the centers of stars, the fact that there is no stable or even remotely stable atomic nucleus with an atomic weight of 5 (that is, 3 protons and 2 neutrons, or 3 neutrons and 2 protons, or 4 protons and 1 neutron, etc.) stopped the formation of elements at helium. In the Big Bang the conditions were not suitable for three helium nuclei to combine into a carbon nucleus because, by the time helium had started to form in significant quantities, the Universe was too cool to allow the triple alpha reaction. In other words, the only stable elements formed in the Big Bang were hydrogen, deuterium, and helium (tritium is unstable and would have decayed rather quickly).

11. This ignores the effects of what astronomers call the Great Attractor—the Virgo-Coma supercluster of galaxies—toward which the Milky Way and the Local Group of galaxies are falling at about 300 kilometers per second due to the gravitational attraction of this huge grouping of mass.

12. The concept of inflation is extremely difficult to visualize. Basically it is not a physical expansion of the Universe but rather a rapid growth of its entire fabric. Imagine that you are a flat being living on the surface of a balloon on which you have a grid marked in centimeters. As the balloon is inflated, everything on the surface expands, including your grid. You do not feel that you have moved because you are growing at the same rate as everything else, and your centimeter scale appears not to have changed, but to someone viewing from outside, everything on the surface of the balloon has increased dramatically in size.

13. Particle physicists use units of electron volts as a convenient measure of the mass of subatomic particles through the energy equivalent of their mass. One joule of energy is equivalent to 6.2×10^{18} electron volts.

CHAPTER 10. What Is There Outside the Universe?

1. Again, before millions of *Star Trek* fans add me to their personal hate lists, I should remind readers that I am a great fan of *Star Trek* and have a wide selection of *Star Trek* books and videos,

which I greatly enjoy, and am as disappointed as anyone to see the series *Star Trek: Enterprise* canceled, having enjoyed the show more than any *Star Trek* version since the original series. However, notwithstanding the recent discoveries described in my earlier book, *Astronomical Enigmas: Life on Mars, the Star of Bethlehem and Other Milky Way Mysteries* (Baltimore: Johns Hopkins University Press, 2005), the idea of traveling across the Galaxy at Warp 8 to get to a distant star system in a few days is genuine fantasy rather than science fiction. While *Star Trek* does take some major liberties with the physical laws of the Universe, it does also genuinely attempt to be plausible in other senses; in this it differs from a film like *Independence Day*, which has so much scientific nonsense that it gets laughable. (Does anyone seriously believe that a 1990s laptop will be capable of communicating with an alien computer using an alien operating system and infecting it with a computer virus written, presumably, for Windows? Or that a 500-kilometer-wide space ship in low Earth orbit would not raise disastrous tides on our planet even before the aliens come streaming out of their ships shooting at us?)

2. It is fair to say that it was the Lensman books, and first of all, *Galactic Patrol*, the third in the series but the first that I read, that really awoke my interest in science fiction. Despite having been published nearly 60 years ago, these books are as popular as ever and, in a poll of fans, were voted the second best science fiction series of all time after Isaac Asimov's Foundation Trilogy. My own copies of the series of books are now extremely battered from long use!

3. A flavor of the story can be gathered from the fact that General Dwight D. Eisenhower, soon to be the president of the United States of America in the real universe, in the book becomes commander of the Venus Sector of the solar system. This is one of the few examples of comic science fiction or, more accurately, science fantasy (because it imagines things that are far outside the laws of the universe), and it is thoroughly recommended as an entertaining read.

4. This was the fourth episode broadcast in season two of the series, in 1967, although it was the thirty-ninth to be produced overall.

5. Not, by a long way, my favorite Asimov novel, but an original and inventive story.

6. In the Lensman series, Kim Kinnison eventually marries a redheaded woman called Clarissa MacDougall. Alert Smith-watchers have noticed that the maiden name of E. E. Smith's redheaded wife was MacDougall and have suggested that he himself provided the model for Kim Kinnison (http://en.wikipedia.org/wiki/E._E._Smith).

7. The University of California does not give Ph.D.'s lightly, which is why one should take his ideas extremely seriously.

8. Yes, this means that there really is a universe in which the 2000 Florida recount was irrelevant because Ralph Nader won all 50 states by a landslide, even though he lost the District of Columbia to George Bush. There will also be an infinite number of universes in which Nazi Germany won the Second World War and the hooked cross is flying over the White House. There may even be a universe where the Los Angeles Clippers have won the NBA championship for the last ten years, but readers will understand that some universes are more probable than others.

9. Here things get too unreasonably complicated to understand. One of the more unexpected discoveries of modern physics has been the fact that a so-called vacuum is not quiet. So-called

empty space is, on a subatomic level, a boiling mass of inconceivable energy with huge quantities of energy being released and reabsorbed almost instantaneously. This makes even empty space itself unstable on the tiniest scales. For a quite brilliant graphic description, I would recommend Arthur C. Clarke's majestic prose in his novel *The Songs of Distant Earth* (London: Grafton Books, 1986), chap. 9, and the acknowledgments at the end of the book.

10. As the development of the universe depends on small random variations in the state of the universe during the initial inflation, in most cases the change will be minimal and the universe formed will be close to what is the average universe. Statistically, the bigger the variation from average, the less likely it is to happen, so, even though universes that turn out completely different from the norm will form, they will only make up a tiny fraction of the total. Statistically, if most universes are close to average, our own Universe should be one of these totally normal and average ones.

11. Yes, the evil scientists in the parallel universe knew that this would happen and do not care, as they are more worried about their own survival. All ends up happily in the end, but to find out how, you must read the book.

12. Were the electromagnetic force just 4 percent stronger than it is, the Sun would turn into a hypernova and all stable stars in the Universe would be no greater in mass than large planets. That is how close our Universe is to being the parallel universe described in *The Gods Themselves*.

13. A report in 2003 by a group made up of physicists from Japan, Switzerland, and China has hinted at indirect evidence of the existence of magnetic monopoles, although the results are, as yet, unconfirmed. You can find a summary, plus the link to the scientific paper announcing the results, at http://physicsweb.org/articles/news/7/10/2. The mass of approximately 10^{16}GeV works out as approximately 0.02 micrograms (0.02 millionths of a gram). This sounds like a tiny mass but, if we remember that a proton has a mass 16 orders of magnitude smaller (that's adding 16 zeros after the decimal point), it gives us an idea of what we are dealing with.

14. A somewhat extreme example of this would be an administration that decides to commission a poll to see if the public supports a particular policy. The equivalent of the anthropic principle would be, instead of asking 1,000 citizens, picked totally at random, to ask instead 1,000 card-carrying, unconditional supporters of the president of the day—that is skewed data.

15. The bigger the object, the smaller the probability of a significant deviation. To choose a somewhat unusual example, quantum physics states that it is *possible* for your clothes to jump spontaneously off your body. However, thankfully for your modesty, they are such a complex system that the quantum leap required for this to happen is vanishingly unlikely. In an infinite universe, though, this and stranger things will eventually happen. You have been warned.

16. If you do not know this particular BBC radio series then shame on you (a degree of absolution may be obtained by following these two links: www.bbc.co.uk/radio4/hitchhikers/ and www.bbc.co.uk/bbc7/drama/progpages/hitchhikers.shtml)! It became such cult listening during my time at university that I started to wonder seriously if prospective students had to demonstrate a thorough knowledge of the series to be admitted to the Physics Department.

17. This was proposed in an essay by E. P. Wigner in the 1960s in *Symmetries and Reflections* (Cam-

bridge, Mass.: MIT Press, 1967). Wigner stated that this happy circumstance "borders on the mysterious" and adds that "there is no rational explanation for it."

18. Even if you cannot prove a particular theory in physics, it is acceptable to be able to propose physical tests whereby you can disprove that theory. This allows the option to refute the theory definitively or, alternatively, to be unable to refute it (that is, to establish that it is a valid possibility), even if there will always be the possibility that some new future test might do so.

Index

The letter t *following a page number denotes a table.*

aberration of light, 65

Alcubierre, Miguel, 77

aliens: flying saucers, 110–13; human interaction with, 72, 100–102, 114, fig. 4.3; level of civilization, 113–14; non-carbon-based life forms, 98–100; number of civilizations, 107–8; probability of, 101–7; public interest in, 96; in science fiction, 96–98; similarity to humans, 98–100. *See also* life; UFOs

Alpha Centauri, 7–9, 8*t*

Andromeda Galaxy, 35, 88, 90–92, 124, 127, 133, 148, 207n17

anisotropies, 151, 211n8

antimatter drives, 74, 204n13

Antonio de Diego, José, 46

Armstrong, Mark, 130

Arp, Halton, 120, 159–61, 212n3

Asimov, Isaac, 71, 97, 175–77

astronomy: benefits from, 1–4; future of, 177–78; history of, 40, 62–68, 123–26, 155; public interest in, 4, 40. *See also* stellar astronomy

Baade, Walter, 91, 124–125

barometric pressure, and sky color, 139

BATSE (Burst and Transient Source Experiment), 51–52, 55

Becquerel, Antoine-Henri, 83

Bessell, Friedrich, 7, 66, 137, 194n2

beta decay, 183, fig. 10.2

BETA project, 109

Betelgeuse, 21–22, 70–71

Bethe-Weizsäcker cycle, 14

B^{2}FH paper, 156

Big Bang: elements formed in, 84–85, 167–68, 213n10; evidence for, 164–69, figs. 9.2, 9.3; and limits of universe, 177–80; and multiverses, 180–81, 184–85, 191; naming of, 156, 208n1; photosphere of, 179; questions remaining about, 184–85; red shift of, 179; symmetry of, 151, 171; visibility of, 178–79

Big Bang theory, 118–19, 140; acceptance of, 140, 149–51, 154–61, 170; questions raised by, 155; version of, 140–41

Big Crunch, 121–22, 141, 169

binary stars, 31–34, fig. 2.1

black holes: in binary stars, 32–34, fig. 2.1; in blazars, 45, 48; at center of galaxies, 34–37; danger from, 34; evidence for, 28–37, fig. 2.1; gas jets from, 36–37, 50, 58, 209n7, fig. 2.2; history of research on, 25–34; and interstellar travel, 77; massive, 35–37; origin of, 22, 33–34, 37, 57–58; in our galaxy, 34; properties of, 24–28, 35–36; public interest in, 24; as radio sources, 35, 209n7; resistance to idea of, 25, 27–28; types of, 27; unknown, 34, fig. 2.2; x-ray emissions from, 31–32, fig. 2.1

blazars, 40–45, 48, 50. See also OJ287

BL Lac objects, 42, 199n6

Boltwood, Paul, 46, 48, 202n21

Bondi, Hermann, 119, 156

Boötes Void, 146

Bowles, Tom, 130

Bradley, James, 65–66, 203n6

Brown, Fredric, 176

Bruno, Giordano, 64

Bunsen, Robert, 9

Burbidge, Geoffrey and Margaret, 156, 160, 198n2

Burst and Transient Source Experiment (BATSE), 51–52, 55

Bussard, R. W., 76

Calvin, Melvin, 101–2

Cannon, Annie Jump, 10, 194n6

carbon-based life forms, alternatives to, 98–100

Carrasco, Luis, 43, 200nn14–15

Carter, Jimmy, 111

Casares, Jorge, 32–33

Cassiopeia A, 124–25

Centaurus A, 29, 124–25

Cepheid variables, 69–70, 86, 90–91, 93

CETI programs, 108–10, figs. 6.3, 6.4

Chandrasekhar, Subramanyon, 23

Chandrasekhar limit, 23, 38, 129

Charles, Phil, 32–33, 46

Charles, Prince of Wales, 39–40

chondrules, 17, 195n15

Christianity, and age of Universe, 80–81

Clarke, Arthur C., 71–72, 74–76, 113, 176, 204n10, 214n9

Clarke refractor, 147

closed universe theory, 140–41

clusters technique, 68–69

CMB. See Cosmic Microwave Background

CNO cycle, 14

The Coal Sack, 142–43

COBE (Cosmic Background Explorer) satellite, 151, 168, 181, 211n8

Columbus, Christopher, 2, 64

Coma cluster, 127, fig. 7.2

Comet de Chéseaux, 136–37, 211n1

Comet McNaught, 211n1

comets, organic molecules in, 103–4

continental drift, age of Earth and, 82, 84, 205n1

Copernicus, Nicolas, 158

cosmic egg, 140

Cosmic Microwave Background (CMB), 150–51, fig. 5.5; as evidence of Big Bang, 164, 168, figs. 9.2, 9.3; and multiverses, 181–82, 185–86, 190; significance of, 151–52, 164–65, 179; smoothness of, 150–51, 181–82, 185–86

cosmological principle, 145–46

cosmologies, nonstandard, 158

cosmology, establishment of discipline, 155

Crab Nebula, 29, 124

Crichton, Michael, 176

Critchfield, Charles, 14

critical universe theory, 140

Crux, 142

Curie, Marie, 83, 205n3

Curtis, Heber, 147

Cygnus, 17, 32, 142

Cygnus A, 28–32, 124–25

Cygnus X-1, 31

dark ages, 178

dark energy, 132–33, fig. 7.6

dark matter, 128, 170

deceleration parameter (q_0), 161–62

de Chéseaux, Jean Philippe Loys, 136–37, 143–45

degeneracy, 21, 23

Delta Cephei, 69

de Vaucouleurs, Gerard, 92, fig. 5.4

Dicke, Robert, 150, 211n7

dinosaurs, extinction of, 56

Dirac, Paul, 156, 185

DNA, alien life forms and, 98, 100

double degeneracy model, 23

double stars, supernova of, 22–23

Drake, Frank, 108–9

Driver, Simon, 144

dust in space, and darkness of night sky, 141–43

Eagle Nebula, fig. 1.5

Earth: age of, 12–13, 81–84, 205n1, fig. 5.1; core temperature of, 12–13; gamma ray bursts and, 56; Greek astronomers on, 5–6, 63; internal structure of, 195n16; suitability for life, 104–5

Eddington, Arthur, 13

Ehman, Jerry, 109

Einstein, Albert: and cosmological constant, 90, 118, 132–33, 148, 207n18, 211n6; field equations of, 26–27, 148, 197n4; and gravitational red shift, 198n12; on multiverses, 190; on quantum theory, 187; and speed of light as speed limit, 26, 76, 196n2; theory of relativity, 26–27, 37–38, 90, 155–56, 196n2; as *Time* person of the century, 89

elements, formed in Big Bang, 84–85, 166–67, 213n10. *See also* heavy elements

Elodie spectrograph, 105

escape velocity, 25–26, 120–21

Eta Carinae, 34, 58–59

Europa, potential life on, 104

expansion of Universe: as Big Bang evidence, 165–66; and color of night sky, 142, 146–48; Einstein's Lambda term and, 90, 118, 132–33, 207n18; gravity and, 121, 132–34, 140, 161, 169–70; research on, 88–92, 147–48, 178, figs. 5.2, 5.3, 7.7; resistance to idea of, 155, 157–61. *See also* Hubble constant; inflation, of Universe; q

extraterrestrial life. *See* aliens; life

Faraday, Michael, 3

fine-tuning, 185–86, 189

first light, 179

Fischer, Debra, 106–7

Fleming, Wilhelmina, 10

fossils, and age of earth, 81

Fowler, William, 156

Fraunhofer lines, 8–9

Friedman, Herbert, 28

Galactic Halo, 87

galaxies: black holes at center of, 34–37; discovery of, 146–47; distance to, 90; first formation of, 178; groupings of, 158–60; luminosity of, 126–28, 162–63; motion of, 35, 88–92, figs. 5.2, 5.3; nebulae as, 67–68, 88, 147; number of, 163; quasars and, 160, 199n6, fig. 9.1. *See also* Milky Way Galaxy

Galileo, 64, 203n4

Gamma Draconis, 65

gamma ray bursts (GRBs), 32, 51–59, fig. 3.5

Gamow, George, 118–19, 157, 164

generation ships, 73

Giacconi, Richard, 28

giant elliptical galaxies, 127

Gold, Thomas, 119, 156

Goodricke, David, 69

Gore, John Ellard, 11, 147

Grand Unified Theory, 188, 190

gravity: black hole gradient of, 34; dark energy and, 132–33; and expansion of Universe, 121, 132–34, 140, 161, 169–70. *See also* escape velocity

Great Attractor, 213n11

The Great Wall, 146

Greek astronomers, ancient, 5–6, 62–64, 203n1

Green Bank conference, 101–2, 108

greenhouse effect, and possibility of life, 105

hadrons, 167

Haldeman, Joe, 24

Harrison, Harry, 176

Harvard system, 10, 194n6

Hawking, Stephen, 188

heavy elements: amount in stars, 85–86, 106–7, 156, 205n5; origin of, 16, 85

Heidt, Jochem, 46
Heinlein, Robert, 74
Helmholtz, Hermann von, 12
Henderson, Thomas, 7, 66, 194n2
Hermann, Robert, 164
Herschel, William, 8, 67–68, 143, 147
Hertzsprung, Ejnar, 10
Hewish, Anthony, 30
hibernation, for long space journeys, 74
high-energy astrophysics, 29
High-Z Supernova Search Project, 130–31
hot Jupiters, 106, fig. 6.1
Hoyle, Fred, 30, 103–4, 119, 156–57
H-R diagrams, 10–12, 18, fig. 1
HST Key Project, 93–94
Hubble, Edwin, 89–92, 118, 132–33, 148, 156,
    206n15–207n18, 212n3
Hubble constant, 91–93, 207n20, fig. 5.4
Hubble Deep Field, 210n17
Hubble Space Telescope, 4, 18, 54, 56, 86, 93–94, 131,
    160–61, 177, 210n17, figs. 1.3, 1.5, 2.2, 3.5, 3.6, 7.2, 9.1
Hubble volumes, 180–81
Huchra, John, 92–93
Huggins, William, 15
Humason, Milton, 90–92, 207n19
Hyades Cluster, 69
hypernovas, 57–59

inflation, of Universe, 146, 161, 168, 178, 180,
    184–85, 213n12
infrared light, discovery of, 8
Inquisition, 64
intelligence, probability of evolution, 108
interferometry, 125
interstellar probes, 72
interstellar ramjets, 76

Jansky, Karl, 123, 149
Jeans, James, 13, 198n12
Jupiter, 18, 104, 111, 124

Kelvin, Lord, 12, 82–83
Kepler, Johannes, 64–65, 136, 203n5

Kerr black holes, 27
Kerr-Newman black holes, 27
Kikuchi, Sen, 46
Kirchoff, Gustav, 9
Klinkenburg, Dirk, 137
Knack, Roger, 160
Kubrick, Stanley, 3

Lacertids, 42, 199n5
Lampland, Carl Otto, 87–88
Laplace, Pierre Simon, 25–26
Large Magellanic Cloud, 170
Leavitt, Henrietta, 69–70
Lemaître, Georges, 118, 157
Lemaître-Gamow theory, 119
Leo II galaxy, 127
Lick Observatory Supernova Search (LOSS), 130
life: elsewhere in our solar system, 104–5; origin on
    Earth, 102–4; in other solar systems, 105–7, figs.
    6.1, 6.2; search for, 108–10, figs. 6.3, 6.4; suitabil-
    ity of planets for, 102–7, figs. 6.1, 6.2. See also
    aliens
light: escape velocity of, 26; spectrum, discovery
    of, 8; speed of, as limit, 26, 74–77, 196n2. See also
    polarization of light
light years, 66, 72
Lilley, John, 101
Lockyer, Norman, 9
Lovell, Bernard, 123
Lowell, Percival, 88
Lowell Observatory, 87–88
Lundmark, Knut, 89–90
Lynden-Bell, Donald, 34–35

M, 126–31, fig. 7.1
Magellanic Clouds, 127
magnetic monopoles, 185, 215n13
Main Sequence, 11–12, 85, fig. 1.1
Markarian 205, 160–61, fig. 9.1
Mars, potential life on, 104
mathematics, and structure of Universe, 188–89
matter, distribution in Universe, 181–82
Mayer, Robert, 11–12

Mayor, Michael, 105

McKellar, Andrew, 164

Messier, Charles, 35, 137

Messier 16, 18

META projects, 109

meteorites, 17, 84, 103–4, 195n15

microwave radiation. *See* Cosmic Microwave
  Background

Milky Way Galaxy, fig. 8.2; black holes in, 34; H-R
  diagram of stars in, 18; mapping of, 67–68;
  massive stars in, 57–58; obscured view of,
  142–43; possible civilizations in, 107–8; radio
  emissions from center of, 124; size of, 67–68,
  70–71, 91–92, 126–127

Miller, Stanley, 102

Minkowski, Robert, 124

Moon: age of rocks on, 84; distance to, 63; landing
  on, 4; sky of, 138–39, fig. 8.1; x-ray emissions
  from, 28

multiverses: Big Bang and, 180–81, 184–85, 191;
  levels of, 180–90, fig. 10.1; opposition to, 186,
  189–91; proving existence of, 190; in science
  fiction, 175–77, 186–87, 191; travel to, 191

Naylor, Tim, 32–33

nebulae: classification of, 10; color of, 15, 147,
  195n12, 211n3; as galaxies, 67–68, 88, 147; star
  formation from, 15–19, figs. 1.3–1.6; types, 147

Neptune, distance to, 70

neutrinos, 170

neutron stars, 22, 27, 30–32, 57, 129

Newman, Ezra, 27

Newton, Isaac, 8, 194n3

Niven, Larry, 176

nonstandard cosmologies, 158

Nordström, Gunnar, 27

novae, spectral classification of, 10

nuclear fusion, and star's energy, 13–16

nuclear testing, atmospheric, ban on, 51

oceans: as origin of life, 102–3; origin of water in,
  103; salinity and Earth's age, 81

OJ287, 40–51, 199n4, 200n14, 200n15, figs. 3.1–3.3

Olbers, Heinrich Wilhelm Mathäus, 138

Olbers paradox, 136–38, 141–49

Oort Cloud, 70

open universe theory, 140

Order of the Dolphin, 101, 104, 107–8

Ordovician extinction, 56

organic molecules, origin of, 102–4

Orion, 34, fig. 2.2

Orion Nebula, 15–16, figs. 1.2, 1.3

Oro, Joan, 103

oscillating universe theory, 141, 154, 169–70

Ozma project, 108

parallax technique, 65–68, 86, 193n1 (ch. 1), 194n7

parallel universes. *See* multiverses

Penzias, Arno, 149–50, 164, 211n7

perfect cosmological principle, 145–46

period-luminosity relationship, 69, 86–87, 90, 93

Perrine, Charles, 13, 194n8

Phillips, Mark, 129

physical laws, in alternative universes, 183–86,
  188–90

Pickering, Edward, 9–10

Pioneer probes, 72

Pistol Nebula, 34, fig. 3.6

Pistol Star, 34, 58–59, fig. 3.6

planets: definition of, 105; detection methods,
  105–6; distance to, 64–65; extrasolar, 105–7, figs.
  6.1, 6.2; suitability for life, 102–7, figs. 6.1, 6.2

Plaskett's star, 57

Pleiades Cluster, 19, 69

Polaris, 69

polarization of light: from blazars, 42–43; classification of objects by, 202n29; from gamma ray
  burst source, 56–57; from supernovae, 56–57

Pournelle, Jerry, 176

Poyner, Gary, 48

Prime Directive, 107

Procyon, 37

Project Ozma, 108

proteins, structure of, 99

proton-proton (p-p) chain, 14–15, 156

Proxima Centauri, 70

Ptolemy, 64

Puckett, Tim, 130

pulsars, 22, 30, 38

pulsating variables, and interstellar distance, 69–70

Puppis A, 124

q, 121–22, 121t, 125–34

quarks, 167

quasars: characteristics of, 29, 34–37, 41; discovery
of, 209n8; and galaxies, 160, 199n6, fig. 9.1; red
shifts of, 164–66, 212n7; as term, 40

Queloz, Didier, 105

radar, 3, 110, 123, 209n6

radiation, and interstellar travel, 75–76

radioactive elements, 12–13, 17, 83, fig. 5.1

radioactivity, 83–84, 108, 205n2

radio astronomy, 122–26

radio sources in space, 35, 41, 108–10, 124–26,
209n7, figs. 6.3, 6.4

radio sources on Earth, detectability in space, 110

ramjets, interstellar, 76

Reber, Grote, 123, 208n4

red dwarf stars, 10–11, 18, 20, 134, 196n18

red giant stars, 11, 20, 22–23, 213n8

red shift, cosmological: of Big Bang, 179; definition of, 158; evidence for, 166, 212n7; and q,
126–28, 131, fig. 7.1; of quasars, 164–66, 212n7;
resistance to idea of, 158–61, 166; steady state
theory and, 161–65; and Universe expansion, 88,
90–91, 148, fig. 5.3

red shift, gravitational, 198n12

Reissner, Hans, 27

Reissner-Nordström black holes, 27

relativistic beaming, 36, 57–58

relativity, 50, 74–75

relativity, theory of, 26–27, 37–38, 90, 155–56, 196n2

Rincon, Paul, 77

Robotic Optical Transient Search Experiment
(ROTSE), 55–56, fig. 3.4

Römer, Ole, 66

Rosse, Lord, 147

RR Lyrae stars, 69–70, 86, fig. 4.2

Russell, Henry Norris, 10

Rutherford, Ernest, 84

Sagan, Carl, 24, 77, 100–101, 108–9, 176

Salpeter, Edwin, 29, 34–35, 197n6

Sandage, A., 92–93, fig. 5.4

Schrödinger equation, 187

Schwarzschild, Karl, 26–27, 156, 197n3

Schwarzschild, Martin, 156

Schwarzschild black hole, 27

Schwarzschild radius, 26

science fiction, 24, 71–74, 77, 96–98, 175–77,
186–87, 191, 196n2

science fiction writers, characteristics of, 175–76

scientific theory, level of certainty in, 212n5

Scorpius X-1, 28–29

sediments, and age of Earth, 81–82

SERENDIP project, 109

Seti@Home project, 109

Seven Sisters, 19, 196n17

Seyfert, Carl, 198n11

Seyfert galaxies, 36–37, 198n11

silicon-based life forms, 99, 207n2

Sirius, 9, 34, 37

Sirius B, 37, 198n12

61Cygni, 7–8, 8t, 66

sky, color of in daytime, 138–39

sky, darkness of at night, 136–38, 141–49, 151

Slipher, Earl Carl, 87–88

Slipher, Vesto Melvin, 88–91, 118, 147–48, 156,
206n11

Sloan Digital Sky Survey (SDSS), 180–81

Small Magellanic Cloud, 69

Smith, E. E. "Doc," 71, 175, 177

Smith, Francis Graham, 124

solar system: origin of, 17; suitability for life,
104–5

space: flatness of, 185; infinite extent of, 181–82;
subatomic energy levels in, 214n9

space warp bubbles, possibility of, 77

stars: age of, 84–87; color and luminosity of,
7–11, 8t, 12, 19, fig. 1.1; death of, 19–23, 141–42,

144–45; distance to, 7, 61, 65–71, 86–87, fig. 4.1; distribution of, 142, 145–46; energy source of, 11–16, 19–20; formation of, 12, 15–19, 167–68, 178, figs. 1.2–1.6; heavy elements in, 85–86, 106–7, 156, 205n5; lifetimes of, 155, 212n2; number of, 141, 143–44; percentage having planets, 106; size of, 18–19; spectral classification of, 8–11, fig. 1; temperature of, 7–11, fig. 1; travel to, 71–78, 107

steady state theory of Universe, 119–20, 140, 154, 156–57, 161–65

stellar astronomy, 5–15, 23

Struve, Friedrich George, 7, 66

Sun: age of, 13; death of, 19–20; distance to, 63–65; energy of, 11–16, 19–20; formation of, 17; Fraunhofer lines, 8–9; Greek astronomers on, 5–6; lifetime of, 11–12, 16, 85, 133, 155, 212n2; motion of, 91; position in galaxy, 70; radio emissions, 124; size of, 18; x-ray emissions, 28

supergiant stars, 11

super-Jupiters, 105

Supernova Cosmology Project, 130–31

supernovae, figs. 7.3–7.5; and age of Universe, 128–32, 169; cause and process of, 21–23; and new star formation, fig. 1.4; polarization of light from, 56–57; and q, 128–31; red shift and, 166; searches for, 129–31; as term, 198n1; types of, 22–23, 128–29, 166, 196n18, 196n19, 210n12, 210n13

tachyons, 77, 196n2

Takalo, Leo, 40

Tammann, Gustav, 92–93, 129, fig. 5.4

Tegmark, Max, 180–81, 190

telescopes, limitations of, 177–79

Tenerife Experiment, 168, 211n8

Tesla, Nikola, 122–23

Test Ban Treaty (1963), 51–52

Thorne, Kip, 77

time dilation, 75, 158, 166

tired light hypothesis, 143, 158

Titan, potential life on, 104

Tombaugh, Clyde, 88

Triangulum Spiral, 90, 127

triple-alpha reaction, 19–20

Turtledove, Harry, 71, 97, 176

Tycho, 64

UFOs, 71, 97, 110–13, 204n9. See also aliens

Universe: age of, 80–81, 91–94, 131–32, 142, 149, 169, 208n3, fig. 5.5; center of, 145; composition of, 132–33, fig. 7.6; distribution of matter in, 181–82; end of, 117–22, 133–34, 140–41, 169; limits of, 177–80; as mathematical structure, 188–89; measurement of, 61–70; number of stars in, 143–44; size of, 70–71, 91–92, 155; steady state theory of, 119–20, 140; structure of, 146, 170, 182–83, fig. 8.3; visibility of early stages, 178–79. See also expansion of Universe; multiverses

Uranus, discovery of, 67

Urey, Harold, 102

Ursa Major, 91

Ussher, James, 80–81

Valenti, Jeff, 106–7

Valtaoja, Esko, 43, 200n12, 200n14, 200n15

Valtonen, Mauri, 44, 46, 48–50

van den Bergh, Sidney, 92, fig. 5.4

Van Den Broeck, Chris, 77

Vega, 7–8, 8t, 9

Venus, 104, 111, 139

V404 Cygni, 32–34

Vinge, Vernor, 176

Virgo A, 29, 35–36, 124, fig. 2.2

Virgo cluster, 127

Visvanathan, N., 43

Voyager probes, 72

walls, 146

warp drives, 77

Watson-Watt, Robert, 123

weather, and sky color, 139

Wegener, Alfred, 82

Wheeler, John, 188

white dwarf stars, 21, 23, 37–38, 129, 198n12

Wickramasinghe, Chandra, 103–4
Wilkinson Microwave Anisotropy Probe
    (WMAP), 132, 144–45, 169, 181, fig. 5.5
Wilson, Bob, 150, 164
Wolf-Rayet stars, 58
Woolaston, William, 8

x-ray astronomy, 28
x-ray emissions: sources of, in space, 28–32, fig. 2.1;
    from Sun, 28
x-ray spectrography, 28, 197n5

Zwicky, Fritz, 198n1

Figure 1.1. From James B. Kaler, *Stars and Their Spectra*, fig. 3.6. © 1989 Cambridge University Press. Reprinted with the permission of Cambridge University Press

Figure 1.2. © Robert Gendler

Figure 1.3. © HST

Figure 1.4. © Steve Mandel, Hidden Valley Observatory

Figure 1.5. © HST

Figure 1.6. © Wil Milan

Figure 2.1. *left:* NASA, CXC; *right:* NASA/CXC/M.Weiss

Figure 2.2. © HST

Figure 3.1. Author

Figure 3.2. Author

Figure 3.3. Author

Figure 3.4. ROTSE

Figure 3.5. © HST

Figure 3.6. © HST

Figure 4.1. From Martin V. Zombeck, *Handbook of Space Astronomy and Astrophysics*, 2nd ed., p. 52. © 1990 Cambridge University Press. Reprinted with the permission of Cambridge University Press

Figure 4.2. With permission Michael Richmond. This work is licensed under a *Creative Commons License*

Figure 4.3. NASA

Figure 5.1. Author

Figure 5.2. © PNAS

Figure 5.3. © HST

Figure 5.4. © John Huchra, Harvard-Smithsonian Center for Astrophysics

Figure 5.5. NASA/WMAP

Figure 6.1. Rice Genome Newsletter

Figure 6.2. John Whatmough

Figure 6.3. Particle Physics & Astronomy Research Council

Figure 6.4. SETI Institute

Figure 6.5. Jerry R. Ehman, Ohio State University Radio Observatory and the North American AstroPhysical Observatory © 1977–2006

Figure 7.1. From Michael Rowan-Robinson, *Cosmology*, Oxford Physics Series, 2nd ed. (1981), fig. 7.4. By permission of Oxford University Press

Figure 7.2. HST

Figure 7.3. Bruno Leibundgut

Figure 7.4. John L. Tonry, Institute for Astronomy

Figure 7.5. HST

Figure 7.6. NASA/WMAP Science Team

Figure 7.7. NASA/WMAP Science Team

Figure 8.1. NASA

Figure 8.2. Lund Panorama of the Milky Way, Sweden

Figure 8.3. Steve Maddox

Figure 9.1. NASA and the Hubble Heritage Team (STScI/AURA)/R. Knacke (Penn State Erie)

Figure 9.2. NASA/WMAP Science Team

Figure 9.3. NASA/WMAP Science Team

Figure 10.1. Max Tegmark, Massachusetts Institute of Technology

Figure 10.2. Author